THE NATIONAL ACADEMIES
Advisers to the Nation on Science, Engineering, and Medicine

The **National Academy of Sciences** is a private, nonprofit, self-perpetuating society of distinguished scholars engaged in scientific and engineering research, dedicated to the furtherance of science and technology and to their use for the general welfare. Upon the authority of the charter granted to it by the Congress in 1863, the Academy has a mandate that requires it to advise the federal government on scientific and technical matters. Dr. Bruce M. Alberts is president of the National Academy of Sciences.

The **National Academy of Engineering** was established in 1964, under the charter of the National Academy of Sciences, as a parallel organization of outstanding engineers. It is autonomous in its administration and in the selection of its members, sharing with the National Academy of Sciences the responsibility for advising the federal government. The National Academy of Engineering also sponsors engineering programs aimed at meeting national needs, encourages education and research, and recognizes the superior achievements of engineers. Dr. Wm. A. Wulf is president of the National Academy of Engineering.

The **Institute of Medicine** was established in 1970 by the National Academy of Sciences to secure the services of eminent members of appropriate professions in the examination of policy matters pertaining to the health of the public. The Institute acts under the responsibility given to the National Academy of Sciences by its congressional charter to be an adviser to the federal government and, upon its own initiative, to identify issues of medical care, research, and education. Dr. Harvey V. Fineberg is president of the Institute of Medicine.

The **National Research Council** was organized by the National Academy of Sciences in 1916 to associate the broad community of science and technology with the Academy's purposes of furthering knowledge and advising the federal government. Functioning in accordance with general policies determined by the Academy, the Council has become the principal operating agency of both the National Academy of Sciences and the National Academy of Engineering in providing services to the government, the public, and the scientific and engineering communities. The Council is administered jointly by both Academies and the Institute of Medicine. Dr. Bruce M. Alberts and Dr. Wm. A. Wulf are chair and vice chair, respectively, of the National Research Council.

www.national-academies.org

SHARING PUBLICATION-RELATED DATA AND MATERIALS

RESPONSIBILITIES OF AUTHORSHIP IN THE LIFE SCIENCES

Committee on Responsibilities of Authorship in the Biological Sciences

Board on Life Sciences
Division on Earth and Life Studies

NATIONAL RESEARCH COUNCIL
OF THE NATIONAL ACADEMIES

THE NATIONAL ACADEMIES PRESS
Washington, D.C.
www.nap.edu

THE NATIONAL ACADEMIES PRESS 500 Fifth Street, NW Washington, D.C. 20001

NOTICE: The project that is the subject of this report was approved by the Governing Board of the National Research Council, whose members are drawn from the councils of the National Academy of Sciences, the National Academy of Engineering, and the Institute of Medicine. The members of the committee responsible for the report were chosen for their special competences and with regard for appropriate balance.

This study was supported by Contract no. N01-OD-4-2139, Task Order #88 between the National Academy of Sciences and the Department of Health and Human Services/ the National Institutes of Health; Grant No. DBI-0127703 between the National Academy of Sciences and the National Science Foundation; Agreement No. B2001-47 between the National Academy of Sciences and the Sloan Foundation; and the National Research Council Fund. Any opinions, findings, conclusions, or recommendations expressed in this publication are those of the author(s) and do not necessarily reflect the views of the organizations or agencies that provided support for the project.

International Standard Book Number 0-309-08859-3

Additional copies of this report are available from the Board on Life Sciences, 500 Fifth Street, N.W., Washington, D.C. 20001; (202) 334-2236, or the National Academies Press, 500 Fifth Street, N.W., Lockbox 285, Washington, D.C. 20055; (800) 624-6242 or (202) 334-3313 (in the Washington metropolitan area); Internet, http://www.nap.edu

Cover: Details from the library ceiling of the National Academy of Sciences building (Lee Lawrie, sculptor)
Front cover: Recording of discovery
Back cover: Reading of the record

Copyright 2003 by the National Academy of Sciences. All rights reserved.

Printed in the United States of America

COMMITTEE ON RESPONSIBILITIES OF AUTHORSHIP IN THE BIOLOGICAL SCIENCES

THOMAS R. CECH (*Chair*), Howard Hughes Medical Institute, Chevy Chase, Maryland
SEAN R. EDDY, Howard Hughes Medical Institute; Washington University, St. Louis, Missouri
DAVID EISENBERG, Howard Hughes Medical Institute; University of California, Los Angeles
KAREN HERSEY, Massachusetts Institute of Technology, Cambridge
STEVEN H. HOLTZMAN, Infinity Pharmaceuticals, Inc., Boston, Massachusetts
GEORGE H. POSTE, Health Technology Networks, Gilbertsville, Pennsylvania
NATASHA V. RAIKHEL, University of California, Riverside
RICHARD H. SCHELLER, Genentech, Inc., South San Francisco, California
DAVID B. SINGER, GeneSoft, Inc., South San Francisco, California
MARY C. WALTHAM, Independent Publishing Consultant, Princeton, New Jersey

Project Staff

ROBIN A. SCHOEN, Study Director
BRIDGET K. B. AVILA, Senior Project Assistant
ELIA BEN-ARI, Science Writer
NORMAN GROSSBLATT, Editor

BOARD ON LIFE SCIENCES

COREY S. GOODMAN (*Chair*) University of California, Berkeley, California
R. ALTA CHARO, University of Wisconsin, Madison, Wisconsin
JOANNE CHORY, The Salk Institute for Biological Studies, La Jolla, California
DAVID J. GALAS, Keck Graduate Institute of Applied Life Science, Claremont, California
BARBARA GASTEL, Texas A&M University, College Station, Texas
JAMES M. GENTILE, Hope College, Holland, Michigan
LINDA GREER, Natural Resources Defense Council, Washington, District of Columbia
ED HARLOW, Harvard Medical School, Cambridge, Massachusetts
ELLIOT M. MEYEROWITZ, California Institute of Technology, Pasadena, California
ROBERT T. PAINE, University of Washington, Seattle, Washington
GREGORY A. PETSKO, Brandeis University, Waltham, Massachusetts
STUART L. PIMM, Columbia University, New York, New York
JOAN B. ROSE, University of South Florida, St. Petersburg, Florida
GERALD M. RUBIN, Howard Hughes Medical Institute, Chevy Chase, Maryland
BARBARA A. SCHAAL, Washington University, St. Louis, Missouri
RAYMOND L. WHITE, DNA Sciences, Inc., Fremont, California

Senior Staff

FRANCES SHARPLES, Director

Preface

I agreed to chair the National Research Council's Committee on Responsibilities of Authorship in the Biological Sciences because I thought the topic is central to the proper conduct of research. And it is an important topic to revisit now because genome databases and other large datasets have greatly ramped up the value of "published materials" while the increasing entanglement of academic and commercial research has complicated the landscape on which science is pursued. I also thought it would be a relatively easy task: after all, isn't there a consensus that publication-related data and materials need to be freely shared?

Now, more than a year later, it is clear to me and the committee that there is in fact a general consensus about sharing published data and materials, but also wide variation in how this implicit contract to share is implemented and in whether individual scientists, companies, or editors exempt themselves in particular circumstances. One hears academic scientists explain, "We always send out our transgenic mice after we publish . . . but of course we expect to be coauthors on any publications that result." One hears company scientists proclaim adherence to the same principle of sharing, "but of course you first need to sign an agreement granting us an exclusive license to commercialize any discovery made with our database or materials." Thus, as in many human activities, the devil is in the details. As a result, the committee ended up not simply recording the community standards as they are practiced, but gleaning from them principles and recommendations that we think are worth adopting generally.

The process the committee traversed in its deliberations is prescribed by the National Research Council to maximize fairness. There was even a meeting in which the Committee was asked whether it had broad enough representation; we decided we did not, and additional industrial representatives were recruited. A public meeting held at the National Academy of Sciences drew a large and diverse audience whose opinions were taken into account. As drafts of the report were written, the committee's deliberations intensified. I had anticipated that there would sometimes be differences of opinion between academic and industrial members; to my surprise, there was no such divide: everyone on the committee felt strongly that once they publish, academic and company scientists take on the same responsibilities to share and should enjoy the same benefits of receiving published materials, data, and software. Finally, detailed anonymous critiques from a diverse group of reviewers led to useful modifications and inclusion of more examples in the report.

The question the committee heard over and over again was, "Shouldn't there be exceptions to the general responsibility to share?" We therefore devote an entire chapter to analysis of such questions. While there are some obvious justifications for exceptions—for example, if it is illegal for a scientist from a particular country to send out a particular type of material—in general, the committee held to *a uniform principle for sharing integral data and materials expeditiously*, or UPSIDE. The upside of UPSIDE is two-fold: it keeps science honest, and it fosters the progress of science. Both are worth nurturing and protecting.

Thomas R. Cech
Chairman

Acknowledgment of Reviewers

This report has been reviewed in draft form by individuals chosen for their diverse perspectives and technical expertise, in accordance with procedures approved by the NRC's Report Review Committee. The purpose of this independent review is to provide candid and critical comments that will assist the institution in making its published report as sound as possible and to ensure that the report meets institutional standards for objectivity, evidence, and responsiveness to the study charge. The review comments and draft manuscript remain confidential to protect the integrity of the deliberative process. We wish to thank the following individuals for their review of this report:

Paul Evans, Ohio State University, Columbus, Ohio; *Journal of Money, Credit, and Banking*, Columbus, Ohio
Philip Campbell, *Nature*, London, England, United Kingdom
Kevin Davies, *Bio-IT World*, Framingham, Massachusetts
Maria Friere, The Global Alliance for Tuberculosis Drug Development, New York, New York
W.R. "Reg" Gomes, Division of Agriculture and Natural Resources, University of California, Oakland, California
Donald Kennedy, Stanford University, Stanford, California; *Science* Magazine, Washington, D.C.
David Korn, Association of American Medical Colleges, Washington, D.C.

Tom E. Lovejoy, H. John Heinz III Center for Science, Economics, and the Environment, Washington, D.C.
Andrew Neighbour, University of California, Los Angeles, California
Peter H. Raven, Missouri Botanical Garden, St. Louis, Missouri
Joseph V. Smith, The University of Chicago, Chicago, Illinois
Oliver Smithies, University of North Carolina, Chapel Hill, North Carolina
Philip P. Green, University of Washington School of Medicine, Seattle, Washington
Randy Scott, Genomic Health, Inc., Redwood City, California
Lincoln Stein, Cold Spring Harbor Laboratory, Long Island, New York

Although the reviewers listed above have provided constructive comments and suggestions, they were not asked to endorse the conclusions or recommendations nor did they see the final draft of the report before its release. The review of this report was overseen by Gilbert S. Omenn of the University of Michigan and C. H. "Herb" Ward of Rice University. Appointed by the National Research Council, they were responsible for making certain that an independent examination of this report was carried out in accordance with institutional procedures and that all review comments were carefully considered. Responsibility for the final content of this report rests entirely with the authoring committee and the institution.

Acknowledgments

This report is the product of many individuals. In particular, we would like to thank all those who attended our workshop, Community Standards for Publication-Related Data and Materials, on February 25, 2002. Without the input of each of these participants, this report would not have been possible.

Mark Adams, Celera Genomics
Wendy Baldwin, National Institutes of Health
Catherine Ball, National Science Foundation
Jules Berman, National Cancer Institute
Helen Berman, Rutgers University
Steven Briggs, Torrey Mesa Research Institute
Eric Campbell, Harvard University
Phil Campbell, *Nature*
Michelle Cimbala, Sterne, Kessler, Goldstein, and Fox, PLLC
Barbara Cohen, *The Journal of Clinical Investigation*
Francis Collins, National Human Genome Research Institute
Katie Cottingham, *Science* Magazine
Nicholas Cozzarelli, University of California-Berkeley,
    Proceedings of the National Academy of Sciences
Jeffrey Drazen, *The New England Journal of Medicine*
Anita Eisenstadt, National Science Foundation
Lila Feisee, Biotechnology Industry Organization
Maria Freire, The Global Alliance for Tuberculosis Drug
    Development

Elisabeth Gantt, University of Maryland
Barbara Gastel, Texas A&M University
Michael Gazzaniga, Dartmouth College
Corey Goodman, Renovis, Inc.
Laurie Goodman, *Genome Research*
Robert Haselkorn, The University of Chicago
Michael Hayden, University of British Columbia
Kathy Hudson, National Human Genome Research Institute
Barbara Jasny, *Science* Magazine
Elke Jordan, National Human Genome Research Institute
Donald Kennedy, Stanford University, *Science* Magazine
Carter Kimsey, National Science Foundation
Marc Kirschner, Harvard Medical School
Stephen Koslow, National Institute of Mental Health
Enno Krebbers, DuPont, University of Delaware
David Kulp, Affymetrix
Eric Lander, Whitehead Institute, MIT
Robert Last, Cereon Genomics
Eaton Lattman, Johns Hopkins University
Craig Liddell, Paradigm Genetics
Ann Link, American Association of Immunologists
Karin Lohman, Committee on Science, United States House of Representatives
Pal Maliga, Waksman Institute, Rutgers University
Cheryl Marks, National Cancer Institute
Victoria McGovern, Burroughs Wellcome Fund
Ira Mellman, Yale University School of Medicine
Joachim Messing, Waksman Institute, Rutgers University
Kate Murashige, Morrison & Foerster, LLP
Elizabeth Neufeld, University of California-Los Angeles, School of Medicine
Ari Patrinos, U.S. Department of Energy
Jerome Reichman, Duke University Law School

Acknowledgments

Ellis Rubenstein, *Science* Magazine
James Siedow, Duke University
Vivian Siegel, *Cell*
Jane Silverthorn, National Science Foundation
Fintan Steele, *Molecular Therapy*
Diane Sullenberger, *Proceedings of the National Academy of Sciences*
Herbert Tabor, National Institute of Diabetes & Digestive & Kidney Diseases
Heidi Wagner, Genentech, Inc.
Bob Waterston, Washington University School of Medicine
Jim Wells, Sunesis Pharmaceutical, Inc.
Sandra Wolman, Universities Associated for Research and Education in Pathology

Contents

EXECUTIVE SUMMARY	1
1 STUDY OVERVIEW AND BACKGROUND	17
2 THE PURPOSE OF PUBLICATION AND RESPONSIBILITIES FOR SHARING	27
3 SHARING DATA AND SOFTWARE	35
4 SHARING MATERIALS INTEGRAL TO PUBLISHED FINDINGS	51
5 DIFFERENT INTERPRETATIONS OF EXISTING STANDARDS	61
6 ENCOURAGING COMPLIANCE WITH AND CONTINUING THE DEVELOPMENT OF STANDARDS	69
REFERENCES	79
APPENDIXES:	
A COMMITTEE BIOGRAPHIES	81
B WORKSHOP AGENDA AND SITUATIONS	89

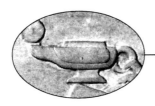

Executive Summary

The publication of experimental results and sharing of research materials related to those results have long been key elements of the life sciences. Over time, standard practices have emerged from communities of life scientists to facilitate the presentation and sharing of different types of data and materials. But recently there is a concern that, in practice, publication-related data and materials are not always readily available to the research community. Moreover, in some fields questions have arisen about whether standard practices really exist, or whether putative standards are accepted by and commonly applied to all authors.

That uncertainty is driven by several factors, including the changing nature of the participants in the scientific enterprise, the growing role of large datasets in biology, the cost and time involved in producing some data and materials, and the commercial and other interests of authors in their research data and materials. These circumstances have engendered widespread interest in a reevaluation of the responsibilities of authors to share publication-related data and materials.

As interest in the topic of standard practices was growing, the National Academies approached the National Cancer Institute, National Human Genome Research Institute, National Science Foundation, and the Sloan Foundation with the idea of undertaking a study of the issues related to sharing publication-related data and materials. With their support, in October 2001, the Academies created the Committee on

Responsibilities of Authorship in the Biological Sciences, whose members were chosen from academe and the commercial sector for their expertise in the life sciences and medicine, and their experience with issues related to scientific publishing, databases, software, intellectual property rights, and technology transfer. The committee was given the following charge:

> To conduct a study to evaluate the responsibilities of authors of scientific papers in the life sciences to share data and materials referenced in their publications. The study will examine requirements imposed on authors by journals, identify common practices in the community, and explore whether a single set of accepted standards for sharing exists. The study will also explore whether more appropriate standards should be developed, including what principles should underlie them and what rationale there might be for allowing exceptions to them.

To meet its charge and obtain a variety of perspectives on these issues, the committee organized a workshop, "Community Standards for Sharing Publication-Related Data and Materials," that was held on February 25, 2002 at the National Academy of Sciences in Washington, DC. The participants included distinguished members of the life-sciences community—researchers and administrators from universities, federal agencies, and private industry; scientific-journal editors; and members of the legal and university technology-transfer communities. Evaluation of the issues was stimulated by the group's analysis of several hypothetical situations (attached in an appendix to the full report) that captured many of the difficult issues facing the community.

During the workshop, discussions about which data and materials related to a publication an author ought to provide and the precise manner in which they should be shared with others revealed how important those requirements are to the scientific community. Much of the analysis that took place in working groups was an effort to discern how an author (with individual competitive, commercial, or other interests) could, by some minimum effort, meet the collective needs of the community. Regardless of the specifics of the hypothetical problem under discussion, the ability to resolve the situation satisfactorily depended ultimately on whether an author could meet the community's general expectations of getting what was needed to move science forward.

While largely unwritten, the community's expectations of authors are a reflection of the value of the publication process to the life-sciences community. The central role of publication in science also explains its value to scientists who want to publish their findings. For individual investigators, publication is a way of receiving intellectual credit and recognition from one's peers (and perhaps the broader public) for the genesis of new knowledge and the prospect of its conversion into beneficial goods and services. Publication also enhances a researcher's job prospects, ability to be promoted or gain tenure, and prospects for research support.

Companies whose scientists publish their findings also typically receive the intellectual credit, recognition, and prestige that come with such disclosures to the entire scientific community. Such nonfinancial benefits can translate into publicity and increased perceived value of a company to investors and business partners. They also strengthen the scientific reputation of the company in the eyes of potential collaborators, employees, and users of the company's products.

Regardless of the motivation, the arena of publication is where participants in the research enterprise share, and are recognized for, their contributions to science. Ultimately, this system benefits all members of the scientific community and promotes the progress of science. Although society encourages innovation in other ways (for example, through the patent system), the sharing of scientific findings, data, and materials through publication is at the heart of scientific advancement. A robust and high-quality publication process is, therefore, in the public interest.

In this context, and informed by the views expressed at the workshop and its own subsequent deliberations, the committee found that the life-sciences community does possess commonly held ideas and values about the role of publication in the scientific process. Those ideas define the responsibilities of authors and underpin the development of community standards—practices for sharing data, software, and materials adopted by different disciplines of the life sciences to facilitate the use of scientific information and ensure its quality. Central to those ideas is a concept the committee called "the uniform principle for sharing integral data and materials expeditiously (UPSIDE)," as follows:

Community standards for sharing publication-related data and materials should flow from the general principle that the publication of scientific information is intended to move science forward. More specifically, the act of publishing is a *quid pro quo* in which authors receive credit and acknowledgment in exchange for disclosure of their scientific findings. An author's obligation is not only to release data and materials to enable others to verify or replicate published findings (as journals already implicitly or explicitly require) but also to provide them in a form on which other scientists can build with further research. All members of the scientific community—whether working in academia, government, or a commercial enterprise—have equal responsibility for upholding community standards as participants in the publication system, and all should be equally able to derive benefits from it.

In addition to UPSIDE, the committee identified five corollary principles associated with sharing publication-related data, software, and materials. The five principles further elucidate the common expectations of the life-sciences community of an author's responsibilities and form the basis of community standards tailored to the types of data and material integral to a particular field and the unique circumstances of research in a discipline. For example, the gene expression community is developing standards for sharing published microarray data, biological taxonomists are promoting a central repository for morphological images, and specialized distribution centers have arisen for many types of plant germplasm. Given the diversity of disciplinary communities in the life sciences, different standards are expected to arise. Nevertheless, the standards reflect a common basis in the principles identified in this report.

As noted in the full report, however, the details of community standards and the nuances of how the principles that underpin them should be interpreted are sometimes a matter of debate within disciplines. Some of these subtleties are discussed in the full report; the chapter in which they are addressed is indicated next to each of the five principles listed below.

DATA AND SOFTWARE

Principle 1. (Chapter 3) Authors should include in their publications the data, algorithms, or other information that is central or integral to the publication—that is, whatever is necessary to support the major claims of the paper and would enable one skilled in the art to verify or replicate the claims.

This is a *quid pro quo*—in exchange for the credit and acknowledgement that come with publishing in a peer-reviewed journal, authors are expected to provide the information essential to their published findings.

Principle 2. (Chapter 3) If central or integral information cannot be included in the publication for practical reasons (for example, because a dataset is too large), it should be made freely (without restriction on its use for research purposes and at no cost) and readily accessible through other means (for example, on-line). Moreover, when necessary to enable further research, integral information should be made available in a form that enables it to be manipulated, analyzed, and combined with other scientific data.

Because scientific publication is intended to move science forward, an author should provide data in a way that is practical for other investigators. The data might reasonably be provided on-line but should be available on the same basis as if they were in the printed publication (for example, through a direct and open-access link from the paper published on-line). Making data that is central or integral to a paper freely obtainable does not obligate an author to curate and update it. While the published data should remain freely accessible, an author might make available an improved, curated version of the database that is supported by user fees. Alternatively, a value-added database could be licensed commercially.

Principle 3. (Chapter 3) If publicly accessible repositories for data have been agreed on by a community of researchers and are in general use, the relevant data should be deposited in one of these repositories by the time of publication.

The purpose of using publicly accessible data repositories is a practical one—to expedite scientific progress and provide access to data in a manner that allows others to build on it. By their nature, these repositories help define consistent policies of data format and content, as well as accessibility to the scientific community. The pooling of data into a common format is not only for the purpose of consistency and accessibility. It also allows investigators to manipulate and compare datasets, synthesize new datasets, and gain novel insights that advance science.

MATERIALS

Principle 4. (Chapter 4) Authors of scientific publications should anticipate which materials integral to their publications are likely to be requested and should state in the "Materials and Methods" section or elsewhere how to obtain them.

Consistent with the spirit and principles of publication, materials described in a scientific paper should be shared in a way that permits other investigators to replicate the work described in the paper and to build on its findings. If a material transfer agreement (MTA) is required, the URL of a Web site where the MTA can be viewed should be provided. If the authors do not have rights to distribute the material, they should supply contact information for the original source. A frequently requested reagent can be made reasonably available in the commercial market or by an author's laboratory for a modest fee to cover the costs of production, quality control, and shipping.

Principle 5. (Chapter 4) If a material integral to a publication is patented, the provider of the material should make the material available under a license for research use.

When publication-related materials are requested of an author, it is understood that the author provides them (or has placed them in an authorized repository) for the purpose of enabling further *research*. That is true whether the author of a paper and the requestor of the materials are from the academic, public, private not-for-profit, or commercial (for-profit) sector. Notwithstanding legal restrictions on the distribution of some materials, authors have a responsibility to make published materials available to all other investigators on similar, if not identical, terms.

During the workshop, it was recognized that the responsibility for creating, updating, and enforcing community standards for sharing publication-related data and materials lies with all members of the community who participate in the publication process and have an interest in the progress of science. This includes academic, government, and industrial scientists; scientific societies, publishers, and editors of scientific journals; and institutions and organizations that conduct and fund scientific research. In addition to creating, implementing, and enforcing standards, some workshop participants suggested that the scientific community should also confront the problems that contribute to uncertainty surrounding standards, for example by creating incentives to share data and materials, and addressing the costs, administrative barriers, and commercial issues related to sharing.

Reflecting these concerns, the committee developed a set of recommendations that describe possible actions by participants in the scientific enterprise to address issues concerning sharing publication-related data and materials. The committee puts these recommendations forward for further discussion and consideration as best practices by the life-sciences community, whose members have the ultimate responsibility to develop and implement community standards.

Recommendation 1. (Chapter 3) The scientific community should continue to be involved in crafting appropriate terms of any legislation that provides additional database protection.

Some companies have identified the lack of commercial protection for databases as the key reason why they need to require investigators who want publication-related data to sign an agreement about their use of the data with the company. Database protection is important to the publication process because it could affect how and whether the community can use and recombine data held in databases. In the past, legislative proposals for increased database protection have been perceived by the community as having potentially negative consequences for sharing and using scientific data. It is in the interest of the life-sciences community to be an active participant in ensuring that any proposed database protection is consistent with the principles of publication and enables researchers working in companies to publish on the same terms as other authors.

Recommendation 2. (Chapter 4) It is appropriate for scientific reviewers of a paper submitted for publication to help identify materials that are integral to the publication and likely to be requested by others and to point out cases in which authors need to provide additional instructions on obtaining them.

Most journals today explicitly or implicitly require that authors provide enough detail about their materials and methods to allow a qualified reader to verify, replicate, or refute the findings reported in a paper. Members of the scientific community support the publishing process by participating as peer-reviewers, often requesting additional supporting information. Identifying materials likely to be requested is consistent with that practice.

Recommendation 3. (Chapter 4) It is not acceptable for the provider of a publication-related material to demand an exclusive license to commercialize a new substance that a recipient makes with the provider's material or to require collaboration or coauthorship of future publications.

Authors should enable others to build on their findings. To build on the author's work, a recipient might need to assemble materials from multiple providers, and they cannot all be granted exclusive licenses. Demanding an *exclusive* license to a new substance made by another investigator using the author's material will effectively block the recipient from assembling the materials needed to conduct research. In addition, although collaborations and coauthorship often arise naturally when materials are shared (to the mutual benefit of the scientists involved) it is unacceptable to require collaboration or coauthorship as a condition of providing a published material, because that requirement can inhibit a scientist from publishing findings that are contrary to the provider's published conclusions.

Recommendation 4. (Chapter 4) The merits of adopting a standard MTA should be examined closely by all institutions engaged in technology transfer, and efforts to streamline the process should be championed at the highest levels of universities, private research centers, and commercial enterprises.

The purpose of sharing publication-related materials is to enable research—that is, to allow the recipients of material to replicate and build on the work of the authors—and the terms of MTAs and their negotiation should not create a barrier to this goal. Because there are so many nuances in the negotiation of MTA-related issues, there is a potential for delay in reaching agreement, and sometimes there is an impasse. The proliferation of MTAs with idiosyncratic requirements set by multiple institutions is, in the end, an impediment to sharing publication-related materials.

Recommendation 5. (Chapter 4) As a best practice, participants in the publication process should commit to a limit of 60 days to complete the negotiation of publication-related MTAs and transmit the requested materials or data.

Such a commitment would eliminate uncertainty for the requestors of materials and remove what is currently perceived as a substantial barrier to the ability of investigators to move forward with their research plans. If sharing publication-related materials in a timely fashion is important to participants in the publication process, authors and others should encourage their institutions to commit to achieving that goal.

Recommendation 6. (Chapter 6) Scientific journals should clearly and prominently state (in the instructions for authors and on their Web sites) their policies for distribution of publication-related materials, data, and other information. Policies for sharing materials should include requirements for depositing materials in an appropriate repository. Policies for data sharing should include requirements for deposition of complex datasets in appropriate databases and for the sharing of software and algorithms integral to the findings being reported. The policies should also clearly state the consequences for authors who do not adhere to the policies and the procedure for registering complaints about noncompliance.

Many journals do not specify policies about sharing data and materials in their instructions to authors. By incorporating transparent standards into their official policies (including a statement of consequences for authors who do not comply), journals can encourage compliance. It is not known how many instances of noncompliance are ever brought to the attention of journal editors or other external authorities; however, a letter from the editor-in-chief or managing editor is often sufficient to resolve problems. Although some journal editors would consider denying a noncomplying author further rights to publish in their journals, on rare occasions, public opinion might be the most influential way to obtain an author's compliance. A journal might choose to declare an author's noncompliance (after all honest attempts were exhausted) in a specific section dedicated to this purpose.

Recommendation 7. (Chapter 6) Sponsors of research and research institutions should clearly and prominently state their policies for distribution of publication-related materials and data by their grant or contract recipients or employees.

The National Science Foundation, National Institutes of Health (NIH), and other funding organizations, such as the Howard Hughes Medical Institute, have policies that reinforce and in some cases extend the standards set by the research community for depositing data in public databases. NIH has also issued a set of principles and guidelines on obtaining and disseminating biomedical research resources, although these are not tied specifically to publication. Universities and private sector sponsors should consider adopting policies that facilitate the distribution of publication-related data and materials.

Recommendation 8. (Chapter 6) If an author does not comply with a request for data or materials in a reasonable time period (60 days) and the requestor has contacted the author to determine if extenuating circumstances (travel, sabbatical, or other reasons) may have caused the delay, it is acceptable for the requestor to contact the journal in which the paper was published. If that course of action is not successful in due course (another 30 days), the requestor may reasonably contact the author's university or other institution or the funder of the research in question for assistance. Those entities should have a policy and process in place for responding to such requests for assistance in obtaining publication-related data or materials.

Few universities, research institutions, or funding organizations have published procedures for resolving problems of noncompliance by their employees or grantees. Although a telephone call to an author from a program director or other representative of an organization can be effective in achieving compliance, funding organizations and research institutions, like journals, can encourage compliance earlier in the process by developing and enforcing transparent policies that encourage sharing of research resources.

Recommendation 9. (Chapter 6) Funding organizations should provide the recipients of research grants and contracts with the financial resources needed to support dissemination of publication-related data and materials.

One reason that researchers have cited for not sharing published materials is the time, effort, and cost involved in doing so. This is a legitimate concern that research sponsors should address. By supporting the development of repositories, allowing grantee institutions to recoup the costs of distribution, and through other mechanisms, funding organizations can help to assist scientists in meeting their obligations as authors. Authors should take advantage of existing ways to facilitate and minimize the costs of sharing publication-related research resources, including the deposition of research materials in existing public repositories. Some researchers have established their own "cottage industries" for producing and distributing commonly requested materials.

Recommendation 10. (Chapter 6) Authors who have received data or materials from other investigators should acknowledge such contributions appropriately.

Authors often fail to acknowledge those who have provided materials, data, or other information that helped in obtaining the findings they are publishing. Sharing should be recognized by citing a relevant publication of the donor of the material, and in the acknowledgement section of a paper. Another idea is to create a public database for acknowledgments. Such approaches would make it easier to recognize and reward those researchers who have been generous in sharing publication-related materials, data, software, or other information.

Executive Summary

Community standards are not federal regulations; rather, they are self-imposed by members of the community and are sometimes incorporated in the official policies of journals. During its deliberations the committee became convinced that most arguments for making exceptions to standards could not be rationalized without sacrificing the integrity of the principles of publication. Such arguments include making exceptions to accommodate commercial interests, the original costs of producing data and materials, the vulnerability of young investigators to competition, the existence of contractual agreements with industrial sponsors, and an investigator's right to mine his or her data before others. In considering these arguments, however, the committee found that participants in the publication system were just as likely to benefit as to be hurt by sharing their data and materials. In some instances, avenues other than publication are available for investigators who want to publicize their findings while maintaining control of the related data. In other cases, reasonable and innovative ways can be found to overcome the problems of costs, contractual restrictions, and competition.

At the same time, it is expected that community standards respect laws that protect human subjects or restrict access to radioisotopes, explosives, controlled substances, and certain pathogens. The expectation that an author share publication-related materials is superseded, for example, by prohibitions imposed by many nations on the distribution of biological materials and organisms collected in those countries.

Aside from situations such as those, exceptions unfairly penalize the community, which would have otherwise had access to the data, information, or material being withheld. Furthermore, granting a special exception to certain categories or particular researchers is problematic for a variety of reasons, including the difficulty of deciding who qualifies for the exception. Considering that community standards maintain quality and facilitate the work of the community in moving science forward, the committee observed that exceptions are likely to weaken the effectiveness of that process over the long term:

Universal adherence, without exception, to a principle of full disclosure and unrestricted access to data and materials that are central or integral to published findings will promote cooperation and prevent divisiveness in the scientific community, maintain the value and prestige of publication, and promote the progress of science.

In the committee's view, there should be a single scientific community that operates under a single set of principles regarding the pursuit of knowledge. This includes a common ethic with regard to the integrity of the scientific process and a long-held commitment to the validation of concepts by experimentation and later verification or refutation of published observations.

The focus of this report is on the life sciences, but the principles and standards considered in the committee's deliberations are of a fundamental nature. Although different fields have different accepted norms and practices, the committee hopes that its recommendations will be of interest to scientists in general and that they will prompt additional thoughtful discourse and debate in the scientific community at large.

True science thrives best in glass houses, where everyone can look in.

—Max Perutz

The search for Truth is in one way hard and in another way easy.
For it is evident that no one can master it fully nor miss it wholly.
But each adds a little to our knowledge of Nature, and from
all the facts assembled there arises a certain grandeur.

—Aristotle

CHAPTER ONE

Study Overview and Background

INTRODUCTION

Community standards and policies for the deposition and sharing of materials and data associated with published research findings play an important role in the life-sciences community. Concern that adherence to standard practices of the community was eroding in recent years culminated in February 2001 when *Science* published a paper describing the draft sequence of the human genome by researchers at a company, Celera Genomics. Another paper on a draft version of the human genome, assembled by the publicly funded International Human Genome Sequence Consortium, was published at the same time in *Nature* and the sequence data deposited in GenBank, an annotated collection of all publicly available DNA sequences. Although *Science* usually requires authors to deposit DNA sequences that a paper cites in GenBank or one of the affiliated public databases, in this case it allowed Celera to post its sequence data on the company's own Web site, where they were made available to academic researchers, but with restrictions on the amount of data downloadable from the Web site at any one time. The data were also made available to for-profit companies on different terms.

The decision by *Science* (Kennedy and Jasny, 2001) provoked considerable debate in the life-sciences community.

The debate stimulated interest in revisiting the core principles that underlie community standards, the accepted practices for sharing data, software, and materials that are specific to different disciplines of the life sciences. One might presume that community standards were established long ago and are therefore widely recognized and agreed on, given that scientific publication has existed for more than 3 centuries. This is true in some, but not all, areas of biology. For example, in systematic and evolutionary biology, there are certain widely accepted standards that are routinely observed. In many more recent, rapidly expanding fields, this is not the case. Rapid changes in the life sciences in recent years have led to:

- Disagreement and uncertainty about the responsibilities of authors to share data and materials.
- A sense that, in practice, publication-related materials and data are not always readily available to researchers who desire access to them.
- Suggestions that standards for sharing are not being enforced.
- Controversy over seemingly different application of journal policies to different authors.
- Questions about how standards and policies apply to various types of data and materials, such as large databases and software.
- Suggestions that standards for sharing may be in conflict with federal legislation that encourages commercialization of the results of federally funded research.
- The prospect that new legal protections for databases, particularly in Europe, will complicate the development of comprehensive and consistent standards.
- Uncertainty as to whether academic investigators should be treated differently from industry investigators with regard to the provision of access to their publication-related data or materials.

To address these concerns, the National Research Council created, in October 2001, the Committee on Responsibilities of Authorship in the Biological Sciences, whose members were chosen from academe and the commercial sector for their expertise in the life sciences and medicine and their experience with issues related to intellectual property rights, scientific publishing, data, software, technology transfer, and the structure of the scientific enterprise. The committee was given the following charge:

> To conduct a study to evaluate the responsibilities of authors of scientific papers in the life sciences to share data and materials referenced in their publications. The study will examine requirements imposed on authors by journals, identify common practices in the community, and explore whether a single set of accepted standards for sharing exists. The study will also explore whether more appropriate standards should be developed, including what principles should underlie them and what rationale there might be for allowing exceptions to them.

To meet its charge and to obtain input from the breadth of the life-sciences community, the committee organized a workshop, "Community Standards for Sharing Publication-Related Data and Materials," which was held on February 25, 2002, at the National Academy of Sciences in Washington, DC. The workshop was organized around five hypothetical scenarios (see Appendix B) that served as the basis for examining the wide array of complex issues related to authors' responsibilities for sharing data and materials. More than 70 workshop participants—the keynote speaker, invited panelists, and audience members—discussed the issues in plenary sessions and smaller working groups. The participants comprised distinguished members of the life-sciences community, including researchers and administrators from universities, federal agencies, and private industry; scientific-journal editors; and members of the legal and university technology-transfer communities.

Scope of the Study and Structure of the Report

This report presents a synthesis of the discussions at the workshop and the issues considered by the committee in its deliberations. The report puts forward the committee's findings and recommendations on the key issues facing the life-sciences community with regard to sharing of publication-related data and materials. The rest of this chapter provides background on reasons for addressing these issues. Chapter 2 examines the value of publishing scientific findings and the principles related to the publication of scientific findings. Sharing of data and software and of materials related to publication are treated in Chapters 3 and 4, respectively, and Chapter 5 reviews the various arguments advanced regarding the differing interpretations and consequences of existing standards. Chapter 6 addresses compliance with appropriate community standards, including ways to encourage compliance and to handle cases of noncompliance; it also addresses the challenge of forging community standards that have the robustness and flexibility needed to accommodate the rapid change in life-sciences research that is expected to continue.

The scope of the committee's study was restricted to issues that begin to arise when a paper is submitted for publication. The report therefore does not address questions about the sharing of data that are not being published or unrefereed preliminary or raw data posted on public Web sites before formal peer-reviewed publication. The report also does not address the requirements of the Shelby Amendment of the Freedom of Information Act. As emphasized at the workshop by committee chair Thomas Cech, president of the Howard Hughes Medical Institute, the purpose of the workshop was to address "the responsibility of authors with respect to sharing publication-related data and materials."

While the principles and standards identified in this report have broad applicability to various disciplines within the life sciences, the committee did not conduct a comprehensive examination of practices for sharing of data and materials specific to every discipline. Such practices are tailored to the types of data and material in use and by the unique circumstances of the research. For example, in systematic and evolutionary biology, the long established and only acceptable practice for sharing publication-related voucher specimens is to deposit them in public or accessible repositories, often museums, where they are available to everyone. In microbiology, on the other hand, the use of national repositories to share cultures is not uniform; many scientists maintain and distribute cultures from in-house collections. To the extent that there are multiple communities of life scientists rather than a single community, those disciplines have the ultimate responsibility to develop and implement specific standards that extend from the general principles and standards identified in this report. Although the focus of this report is on the life sciences, the principles and standards considered in the committee's deliberations are of a fundamental nature, and the committee hopes that its recommendations will be of interest to scientists in general.

BACKGROUND: WHY IS THERE A PERCEIVED PROBLEM?

The sharing of experimental results and research materials has long been important for the advancement of science and technology. For many years, a spirit of free and open sharing seemingly prevailed among life scientists. However, today's rapidly evolving research environment is producing some growing pains in the life-sciences community.

Among the common perceived problems are the ignoring or denial of requests for materials or data associated with a publication and long delays in honoring such requests. Increasingly, data and materials that *are* shared come with restrictions, such as material transfer agreements (MTAs) that limit how the resources may be used. Although in some fields of biology, such as x-ray crystallography, more data are shared than ever before; in other life-science fields, the unrestricted, unfettered sharing of data and materials, including those related to published research, is thought to be less common than it was some 20 years ago. Although quantitative evidence is difficult to obtain, a recent survey of geneticists and other life scientists at 100 U.S. universities (Campbell et al., 2002) reported that the ideal of free and open sharing is not always being met. Of geneticists who had asked other academic faculty for additional information, data, or materials regarding published research, 47% reported that at least one of their requests had been denied in the preceding 3 years, and 12% of geneticists acknowledged denying a request from another academic researcher themselves. The phenomenon is not peculiar to academic genetics, according to the survey. There are no a priori reasons to suggest that geneticists were more likely than other university life scientists to report having their requests denied or having denied others' requests. Supporting the notion that sharing is becoming less common, 35% of the geneticists said that sharing of data and materials had decreased during the preceding decade, while only 14% said that it had increased.

Commercial Considerations and Other Concerns

Various factors are believed to contribute to the reduction in unrestricted sharing of publication-related data and materials and to new concerns about sharing in the life-sciences research community. One is the growing role of the for-profit sector—such as pharmaceutical, biotechnology, research-tool, and bioinformatics companies—in basic and applied research over the past 2 decades, and the resulting circumstance that increasing amounts of material and data are in private hands. Although scientists who work for the companies typically want to share reagents and information and many companies see value in sharing, the primary responsibility of a company is to its investors. Giving away valuable data and materials without securing some type of intellectual property protection, and without any promise of financial return, can, depending on costs and competitive implications, result in reluctance to share widely.

Biotechnology and pharmaceutical companies are not only conducting basic research in their own laboratories but also are funding the work of researchers in the not-for-profit sector (universities and private not-for-profit research institutions). This intermingling of for-profit private-sector activities with public or not-for-profit interests increases the prospect of potentially conflicting missions that can impede unrestricted sharing as researchers become involved in commercial activities.

Another major contributor to the current climate is the increasing concern among universities and academic life scientists about intellectual

Study Overview and Background

> **BOX 1-1**
> **Technology Commercialization and Sharing Research Tools**
>
> Many researchers believe that the increased use of license agreements and material transfer agreements interfere with the free exchange of publication-related research resources. One school of thought holds that university technology-transfer offices tend to overestimate the potential commercial value of their own researchers' materials, particularly research tools that are unlikely to be commercialized, as opposed to materials that could be used directly as products in their own right (such as diagnostics, drugs, and vaccines) or used as the basis for new services (such as software databases).
>
> In response to such concerns, the National Institutes of Health (NIH) established a Working Group on Research Tools (www.nih.gov/news/researchtools), and in 1999 issued a set of principles and guidelines for sharing biomedical research resources developed with NIH funding (NIH, 1999). For cases in which an invention supported in whole or in part with federal funds is useful primarily as a research tool, NIH says that "inappropriate licensing practices are likely to thwart rather than promote utilization, commercialization and public availability of the invention."

property rights and commercialization. In the United States it stems in large part from the Patent and Trademark Law Amendments Act, commonly known as the Bayh-Dole Act (PL 96-517, 1980), passed by Congress in 1980. The Bayh-Dole Act encourages universities and other not-for-profit research institutions to promote the use, commercialization, and public availability of inventions developed through federally funded research by allowing them to own the rights to patents they obtain on these inventions. That has contributed to more overlap in the interests of the for-profit and not-for-profit research sectors, and in some cases impeded the unrestricted sharing of publication-related data and material as universities and other not-for-profit research institutions have sought the commercial development of and economic returns from their intellectual property (see Box 1-1). According to workshop panelist Michael Hayden, a professor of medical genetics at the University of British Columbia and chief science officer for Xenon Genetics, Inc., "the blurring between the university and the biotechnology companies has become more and more apparent" in Canada as well—in this case as a result of "an implicit understanding" that, in return for increased government funding, "the universities are going to play a bigger role in commercializing that intellectual property and making sure there is economic benefit."

Commercial interests are not the only reason for withholding data and materials. The geneticists surveyed by Campbell and his colleagues cited additional reasons for intentionally withholding information, data, or materials related to their own published research. They include the financial cost of providing the materials or information to others; the need to preserve patient confidentiality; the need to protect the ability of a graduate student, postdoctoral fellow, or junior faculty member to publish follow-up papers; the need to protect one's own ability to publish follow-up papers; and the likelihood that the recipient will never reciprocate. It is not surprising that a reluctance to share is more common in fields in which scientific competitiveness is high.

The Changing Nature of Data

New types of data and materials are also complicating publication-related sharing practices. One factor that has added a new dimension to the scientific landscape and is an increasing source of concern about community standards for sharing is the growing role of large databases and other large datasets in life-sciences research. The rise of "big science" projects, such as the Human Genome Project, and the ever-increasing pace of technology have enabled researchers to collect vast quantities of data faster and faster. The large databases being compiled include genomic databases, microarray-based gene-expression databases, proteomics databases, large-scale databases for comparative genomics, human population-genetics datasets, ecological datasets, and databases resulting from the use of imaging technologies. Because many of the databases can be productively "mined" for a long time and yield many papers, some authors view relinquishing control of them as a stiff penalty in light of the time, cost, and effort needed to create the first publication.

Although the genomics, structural-biology, and clinical-trials communities have established public databases that facilitate the free sharing of data in standardized formats, researchers in other disciplines that are generating large datasets, such as those resulting from brain imaging or gene and protein expression studies, have yet to agree on standards for when and how to share, format, annotate, and curate data.

The time, effort, and expense involved in generating large datasets, databases, and some research materials have been cited as arguments for restricting access to them. In the case of databases, unrestricted sharing is considered especially problematic because U.S. law does not provide intellectual property protection for databases (see Box 3-1). Any enterprise that produces large databases may be reluctant to share it without restrictions on initial publication, inasmuch as doing so may mean giving up a substantial commercial advantage and could enable the wholesale copying of databases by others for commercial purposes.

Other emerging challenges in publication-related sharing arise from practices related to software and algorithms. These are becoming more common as the subject of publications in the life sciences. Software developers have long disagreed about whether the source code needed for a published program or algorithm should be made available to everyone, and life scientists who develop software are no exception. One reason for the debate is that although software can be copyrighted, it can be difficult in practice to prevent someone else from copying and quickly modifying the source code and taking the lead in commercializing it. And some have argued that mandatory sharing of source code prevents universities from exercising their legal right to develop commercial products from federally funded research.

In the workshop's keynote presentation, Eric Lander, director of the Whitehead Institute Center for Genome Research, reviewed some of the many contentions that are shaping the debate over sharing of data and materials associated with publications and the related topic of public-domain resources. "These are hard arguments to weigh in the absence of an intellectual framework for evaluating them," he said. "I think our goal is to step back and ask, 'What is the intellectual framework in which we can parse these arguments?'" The following chapter examines such a framework.

CHAPTER TWO

The Purpose of Publication and Responsibilities for Sharing

THE TRADITION OF SCIENTIFIC PUBLICATION

The roots of scholarly scientific publishing can be traced to 1665, when Henry Oldenburg of the British Royal Society established the *Philosophical Transactions of the Royal Society*. Oldenburg was motivated, in part, by a desire to remove himself as diplomatic interlocutor between the dispersed, independent scientists of the time with whom he communicated individually. The aim of the new publication was to create a public record of original contributions to knowledge and to encourage scientists to "speak" directly to one another. By providing intellectual credit publicly for innovative claims in natural philosophy, the journal encouraged scientists to disclose knowledge that they might otherwise have kept secret.

The *Philosophical Transactions of the Royal Society* created a sense of competition among scientists to be the first to publish a new scientific finding, an incentive that is continued in modern scientific journals. If the journal is a prominent one, publication endows the author with an extra measure of prestige. In addition, as *Cell* editor Vivian Siegel and other workshop participants noted, publications also yield indirect rewards. For example, they affect a researcher's job prospects and ability to be promoted or gain tenure. Publishing a scientific paper can result in fruitful new scientific collaborations, including financially profitable arrangements for authors in academe, as a result of commercial overtures for collaboration or consultancy.

Publishing also holds some risks for an author. Competitors might use results presented in a paper to advance their own research and "scoop" the original author in future publications. The careers of young scientists might be particularly vulnerable to having prospective research "picked off" by others. (However, if a researcher chooses not to publish his or her results or chooses to delay publication, someone else might publish the same findings first and receive the credit.) Another risk associated with publishing is that other researchers will use information presented in a paper to invalidate or question the author's own findings, and publish conflicting results.

Are the benefits and risks of publishing any different for companies whose investigators publish than those for academic scientists? It was pointed out at the workshop that companies whose scientists publish their findings typically receive the intellectual credit, recognition, and prestige that come with such disclosure to the entire scientific community. Such nonfinancial benefits can translate into increased publicity and increased perceived value of a company to potential investors and business partners. They also strengthen the scientific reputation of companies in the eyes of potential collaborators. By encouraging others to use their methods and materials, companies can develop a net of researchers who are extolling and extending the value of the technology that the company has published. Moreover, companies that encourage their investigators to publish are attractive to employees or potential employees who wish to build and maintain their publication record, either in anticipation of someday returning to academe, as a vehicle for facilitating their participation in and recognition by their peer scientific community, or in buttressing their own career prospects within the company.

For a for-profit research entity, publication also carries financial risks. By revealing proprietary data or other trade secrets, publishing may harm a company's competitiveness in the marketplace and thus endanger the return to investors. A competitor might use information disclosed in a scientific paper to develop a competing product or otherwise gain commercial advantage or to discredit the product claims of the company making the disclosure.

While companies whose scientists publish may worry about their competitive edge in the commercial market, researchers in academe worry about gaining a competitive edge in the rewards process and about getting their research grants renewed. Where academics are rewarded by priority, "fame," and career advancement, companies whose investigators publish receive benefits in terms of visibility, public relations, and validation. Although there are different tradeoffs involved in publishing, in practice, researchers from these two worlds often have similar goals and are motivated by common incentives. Their common interests converge in the forum of scientific publication.

PUBLISHING AND COMMUNITY STANDARDS

By facilitating communication between individuals who had worked in isolation from one another, the *Philosophical Transactions of the Royal Society* also contributed to the development of a scientific community. As a result, modern journals do more than simply register the intellectual accomplishments of *individual* scientists; they record a *collective* body of knowledge. Journals are a centerpiece of the scientific enterprise and serve as a focal point for the description of scientific results. Journal articles supply information that helps scientists to develop new hypotheses, and they provide a foundation on which new scientific discoveries and inventions are built. As Eric Lander noted at the workshop, "science is fundamentally a cumulative enterprise. Each new discovery plays the role of one more brick in an edifice." Authors cite previously published papers to make a case for their conclusions that is based on a combination of previously documented scientific evidence and the new information they have gathered. Scientific journals, many established by learned societies, provide a forum for a continuing dialogue of sorts, as authors discuss findings that add new pieces to others' previously published results or announce alternative conclusions to those made by other authors or contradict them. Science moves forward in this way.

Because publication is central to the activity of the scientific community, and consequently, to scientific progress, principles and standards that govern an author's responsibilities related to publication have always been paramount. As the 1992 National Research Council report *Responsible Science* observed, "For centuries scientists have relied on each other, on the self correcting mechanisms intrinsic to the nature of science, and on the traditions of their community to safeguard the integrity of the research process. This approach has been successful largely because of the widespread acknowledgement that science cannot work otherwise, and also because high standards and reputation are important to scientists" (NRC, 1992).

Because standards related to publication are so important to the functioning of the community, calls for the publication system to adapt to the different risks of publication to scientists working in different circumstances are not easily implemented. Chapter 5 addresses specific arguments related to exceptions, but in general, applying a standard to some authors and not others weakens the incentive of distinction that has attracted scientists since Oldenburg's day to publish publicly in a journal. When exceptions to the community standard are sought and granted, there is a danger that the value of publishing is diminished, not only for the author who requests an exception, but for the entire community. Moreover, if the same standard does not apply to all authors, then the community cannot assume that the quality of scientific papers and the information they purport to represent is reliable. That jeopardizes the integrity of the publication system.

That is not to say that publication-related community standards are insensitive to other important societal interests, such as protecting the identities of research subjects. Measures to protect that information do affect how data are reported and made available to other investigators; nevertheless, the community has striven to find ways to maximize the availability of relevant data without compromising privacy.

A current topic of discussion in the scientific community is the possibility that published information in the life sciences will be exploited by bioterrorists. It is too early to say where those discussions will lead, but current community standards abide by regulations on access to some research materials (for example, radioisotopes, explosives, controlled substances, and pathogens) for good reason. If additional safeguards are found to be necessary in providing access to research data and materials, the community must make accommodations for them.

Similarly, the community complies with prohibitions imposed by some nations on the distribution of biological materials and organisms collected in those countries. Biological materials that are paleontological, archeological, or anthropological in nature, and sometimes samples of organisms, may by national law be required to be deposited in the country of origin, and even when material is allowed to be exported, there are often legal restrictions on its subsequent distribution and use. For example, the commercial use of such samples may be prohibited or restricted. Nonetheless, all such material is made fully available for study at the repository, and not normally under the control of the authors who published results derived from studying it. The details of the results of the original study, and images, DNA sequences, and other information derived from the specimens, are also made available.

The principles and standards of scientific publication are also consistent with society's interest in the applications of scientific knowledge and their economic and other benefits. An author who publishes a paper is expected to share materials related to that publication to other scientists for research purposes, but that does not prevent an author from seeking intellectual property rights protection in order to realize the commercial value of those materials. To encourage the disclosure of scientific information, the patent system bestows inventors of a novel, nonobvious, and useful innovation with the right, for a limited time, to prevent others from making or using that innovation, unless licensed to do so. Scientific publication provides no such incentive, but to the contrary, encourages other scientists to use and integrate into new research those things described in a scientific publication. An author who publishes a scientific paper describing a patented process, for example, may have a legal right to prevent others from using it, but the scientific community holds the expectation that an author will make available a license to use that process for research. From a social perspective, the two systems are complementary: patenting fosters the commercialization of ideas;

scientific publication communicates the ideas that build the edifice of science. Scientific publications also influence the issuance of patent rights by defining the landscape of the "prior art" and "obviousness" criteria used in assessing the novelty of putative patent claims.

JOURNAL POLICIES AND COMMUNITY STANDARDS

Journals have their own policies that describe an author's responsibilities related to publication and sharing publication-related data and materials. Publishers of journals include for-profit companies and not-for-profit enterprises, such as university presses, scientific societies, and associations, and each publisher is motivated by the intellectual objectives and fiduciary responsibilities of its own constituencies. Journal editors often compete for papers that increase the impact and standing of their journals in the scientific community and their mass media coverage. On occasion, journal editors have been willing to make exceptions to their usual policies on data sharing in return for the opportunity to publish a paper they believe will be of high impact in the scientific community and, increasingly, in the general public.

The extent to which journals state their policies for the sharing of materials and data is highly variable (Table 2-1). That variability and the diverse nature of journals might suggest that common principles and standards do not exist. But even the stated policies of journals do not capture what are generally recognized as accepted practices and expectations of the community. For example, most journals today explicitly require that authors provide enough detail about their materials and methods to allow a qualified reader to replicate all experimental procedures. A logical, often implicit, extension of that requirement is that authors must make available the data and materials needed for others to verify or refute the findings reported in a paper. Thus, for example, in a paper citing genetic results from one or a series of organisms, voucher specimens should be cited and deposited in an appropriate public repository where the identity of the organisms can be checked by subsequent workers (with the obvious exception of well-known and easily-available

Table 2-1. Policies of 56 Most Frequently Cited Life-Science and Medical Journals

	Percentage of journals				
Type of Policy	All Journals (N=56)	Society or Association Publishers (N=37)	Commercial Publishers (N=19)	Life-Sciences Journals (N=38)	Clinical-Medicine Journals (N=18)
Sharing materials	39 %	30 %	58 %	47 %	22 %
Sharing software	2	0	5	3	0
Depositing data	41	35	58	53	17
Statement of consequences	2	3	0	0	6
Whom to contact	4	3	5	3	6
No policy	45	49	42	32	72

Note: Journals were identified in a search of the Institute for Scientific Information Journal Citation Reports in the life sciences and medicine. The output was sorted by impact factor; review journals were excluded. The policies of the top 56 journals (as found on their Web pages) were the basis for the table. Percentages were rounded to whole numbers.

strains). Insofar that scientific publication is central to the forward progress of the scientific community, it is presumed that an author must provide data and materials in a way that others can build on them. These widely held expectations are not necessarily incorporated in current journal policies.

THE PRINCIPLES OF PUBLICATION

At the workshop and in its deliberations, the committee attempted to distill the community's most basic interests in the process of publication. It found that a majority of the scientific community held common ideas and values about publication and the role it plays in science, and that those ideas have guided the development of community standards that facilitate the use of scientific information and ensure its quality. Central to those ideas is a concept the committee called "the uniform principle for sharing integral data and materials expeditiously (UPSIDE)," as follows:

Community standards for sharing publication-related data and materials should flow from the general principle that the fundamental purpose of publication of scientific information is to move science forward. More specifically, the act of publishing is a *quid pro quo* in which authors receive credit and acknowledgment in exchange for disclosure of their scientific findings. An author's obligation is not only to release data and materials to enable others to verify or replicate published findings (as journals already implicitly or explicitly require) but also to provide them in a form on which other scientists can build with further research. All members of the scientific community—whether working in academia, government, or commercial enterprise—share responsibility for upholding community standards as equal participants in the publication system, and all should be equally able to derive benefits from it.

Along with UPSIDE, five additional principles guide the development and implementation of community standards. Chapters 3 and 4 discuss those principles and the nuances of how they are embodied in examples of community standards for sharing data, software, and materials. New community standards are likely to evolve as science itself changes, but the principles remain a fundamental underpinning of the their development. The principles motivate the creation of standards that maximize the value of scientific findings to the community, because this has proved to be the way that science progresses most rapidly. In addition to the principles of publication, Chapters 3 and 4 include the Committee's recommendations for increasing the effectiveness of community standards for sharing data and materials.

CHAPTER THREE

Sharing Data and Software

Although the fundamental rationale for sharing publication-related data and materials is relatively straightforward and generally accepted in the scientific community, many workshop participants noted that "the devil is in the details" when it comes to moving beyond the general principle and deciding what is reasonable and necessary to provide in a publication. This chapter examines the details of some of the contentious issues related to the sharing of data and software associated with scientific publications.

The committee considered two fundamental questions: What specific information should be provided to fulfill an author's obligation to share publication-related data and materials? Under what terms or conditions and in what form should that information be provided when practical considerations, such as page limits, preclude its inclusion in the publication itself?

WHICH DATA SHOULD BE SHARED?

In the context of a published finding, the information that should be shared and the manner in which it should be made available depend on how central it is to the principal conclusions of the paper and to the ability of others to validate or refute it. An assemblage of data or a database may itself be the *central* finding of a paper—the results themselves or data that would be shown in the key figures of a publication, if space permitted. For example, in a paper announcing the sequencing of an entire genome, the sequence would be a central aspect of the paper. In other cases, the data are *integral* to the findings being reported, that is, necessary to support the major claims of the paper and essential to enable a knowledgeable peer to reproduce and verify the results. In still other cases, the data or a database provides *background* to a publication—that is, not integral to the findings or conclusions being presented, but without which the findings or conclusions could not have been derived. Background information would not be essential for reproducing, verifying, or building on the claims in the paper; it might be considered as background, for instance, because obvious alternative methods or sources of data could be substituted. A corollary to the uniform principle for sharing integral data and materials expeditiously (UPSIDE), therefore, is the principle that all information that is either central or integral to the paper should be made available in a manner that enables its use for replication, verification, and furtherance of the published claims.

The collection and compilation of large and complex assemblages of data—such as gene sequences, microarray data, and images—are increasing in the life sciences. These datasets or databases have become an important resource in many disciplines. That such large datasets cannot be fitted into the printed version of a paper has led to ambiguity about what an author must provide to readers of the journal.

If a large dataset or database is itself the result being reported in a scientific publication or is integral to the paper, it would be appropriate, but is often impractical, to provide all the data in the paper itself. The data might reasonably be provided on-line but should be available on the same basis as though they were in the printed publication (through a direct and open-access link from the paper published on-line). This principle is an extension of UPSIDE.

If the complete dataset or database was used in a publication but is not integral to the conclusions presented, the authors are free to hold the broader data or database as closely as they wish. In this setting, what should be disclosed are the subsets of data needed to verify and reproduce the specific conclusions. Expert judgment must be exercised during the editorial review process to determine whether the information in question is an integral part of the discovery or merely provides background.

Some members of the scientific community might like to have access to every available piece of information that an investigator has collected during the course of his or her research. In some fields, such as genome sequencing, groups of researchers have set up mechanisms for sharing some unpublished data. However, it is generally accepted that a scientist has not only the right but also the obligation to evaluate, organize, and ascertain the reproducibility of data before their dissemination via publication. Therefore, in presenting their final findings, authors are not obliged to provide all the raw or unprocessed data they have generated.

Sharing large datasets or databases that contain information about human subjects presents a special challenge because of the requirement to protect the rights and privacy of people who participate in research studies. Clinical databases might contain details that would permit linkages to identify research participants. The committee recognizes that databases arising from clinical studies or treatment trials must be made available in a manner that complies with applicable standards for protection of human subjects (Department of Health and Human Services, 2001).

Sharing Software and Algorithms

Publications that deal with software or algorithms, like those involving large datasets or databases, are relatively new in the life-sciences literature. There are no consistent, accepted community standards for sharing such information. As with the other standards discussed in this report, the committee considers that those for sharing software and algorithms should be guided by UPSIDE, as enunciated for other categories of publication: that the purpose of publication is to enable other scientists to verify and build on published work, and that all members of the scientific community have equal responsibilities in and benefits from the publication system. As in the case of data and databases, to be consistent with the principles of publication, anything that is central or integral to a paper should be made available in a manner that enables its use for replication, verification, and furtherance of science.

When the central finding of a scientific paper about software is a new algorithm—the equivalent of a new idea for solving a particular problem or achieving some result—the author must provide enough information so that another investigator in the field can reproduce the finding and build on it. One way to accomplish that is to provide in the paper (or on-line) a detailed description of the algorithm and its parameters. Alternatively, if the intricacies of the algorithm make it difficult to describe in a publication, the author could provide an outline of it in the paper and make the source code (the implementation of the algorithm) available to investigators who wish to test it. Either manner of providing the information upholds the spirit of UPSIDE.

A paper that describes a new software package claimed to be useful for investigating specific types of life-science questions presents a slightly different situation. Here, the intended scientific reading audience for the paper is a wider user community, not other computational or mathematical biologists. The author is claiming a result that is a program that biologists can use, not an algorithm that other software experts could implement in their own software. To be consistent with the principles of sharing publication-related materials and data, the author should provide at least an executable file—and ideally, the source code. That access would enable another investigator to verify the claims of the paper—namely the utility of the package for investigating particular questions. Publishing a paper of this nature would not preclude the author from simultaneously copyrighting or patenting the software and making it available for sale, for example, in a commercial version that can be upgraded continuously, contains special features, or includes user support.

Deciding What Constitutes Central and Integral Information: Sample Scenarios

The hypothetical scenarios described below illustrate how data, algorithms, or software might or might not be considered central or integral to a published finding.

- *Gene sequences.* In considering how to determine whether particular DNA sequence data are integral to a scientific paper, workshop participants discussed several hypothetical journal articles about the kangaroo genome. For a paper titled "The Complete Genome Sequence of the Kangaroo," the complete genome is the result of the paper; therefore, the entire genome sequence should be made available as though it were a figure or table in the paper. (Moreover, the authors should provide a means to verify the species of kangaroo sequenced, and the population from which it was derived.) However, if a paper's central claim or result is to report 57 protein kinases found in kangaroos (such as in a publication entitled "Fifty-seven New Protein Kinases from the Kangaroo"), only the sequences of the 57 kinases must be provided, even if the entire genome was sequenced to obtain this result. A paper entitled "A Complete Catalog of the Protein Kinases in the Kangaroo" would have to disclose the whole genome sequence or database that is necessary to verify the claimed conclusion of completeness. In other words, the primary claims being made in a paper help to guide decisions about which data the authors must make available.

- *Databases.* One of the hypothetical scenarios discussed in detail at the workshop (see Appendix B) concerned two publications related to a new, proprietary model of the human heart ("The Virtual Heart") that incorporates an extensive database of experimental data collected by the authors. Paper A provides an overview of the entire Virtual Heart system, including the elements of the database and the underlying software. Paper B describes a specific result in which the Virtual Heart is used to predict an association between heart disease and a particular genetic variant. The database and software would be considered integral to paper A but not to paper B. To meet the principles of publication for paper A, the authors would be required to provide free access to the database and either a sufficient description of the algorithms on which the software is based or the source code. For paper B, the authors would have to provide evidence to support the association being claimed and at least some description of the parameters of the Virtual Heart model that led to the prediction.

- *Software and algorithms.* As in the examples above, a decision on what must be shared for papers involving software or algorithms depends on the claims being made in the paper. For example, a paper titled "KinaseMagick: A Supersensitive Heuristic Program for Identifying Protein Kinases Better than BLAST" would probably be in the category of a software-package announcement. (BLAST is a public resource that allows researchers to scan all publicly available DNA-sequence data for specific sequence homologies.) The software itself is the principal result being announced and therefore considered integral to allowing others to duplicate the claims and should be made available as described above to support the central claim of the paper. A paper titled "An Improved Motif Detection and Alignment Algorithm Used to Detect 57 Kinases in Kangaroo" is likely to be making a claim that the algorithm is novel, important, and necessary to the work. Here the algorithm is integral, but a specific implementation may not be. The algorithm should be described in sufficient detail to reproduce the experiment (a condition that might be satisfied by releasing source code). Finally, a paper entitled "57 New Kinases in the Kangaroo Genome" that happens to use a custom search program but for which essentially identical results could be reproduced by standard methods, such as BLAST, can simply mention in the description of methods that a custom program was used. Neither the software nor the algorithm is central to reproducing the paper's claims, so they are not considered integral and need not be released to others.

Distinctions about what is or is not integral or central to a scientific paper might not always be clear, and there will always be gray areas. In such circumstances, it is the responsibility of the journal editor and those who are reviewing the paper to make the final decision about the author's responsibilities for data sharing. Furthermore, in evaluating the importance of a paper submitted for publication, it is within a reviewer's purview to consider the extent to which the community can build on the paper's findings. The relevance and importance of a paper is diminished when information (software or data) associated with its central findings are not available or are encumbered. Finally, it should be mentioned that an author's obligation to provide the minimum dataset needed to support a paper's findings is not meant to suggest that authors should necessarily parse their research results into multiple papers. A strategy to withhold data in order to publish a series of papers over time runs the risk that other investigators might publish similar data first.

REASONABLE ACCESS: HOW SHOULD INFORMATION BE PROVIDED?

In addition to the issue of what information should be provided, it is important to consider the matter of *how* that information should be made available. In other words, what constitutes "reasonable" access? In the case of software, several mechanisms exist by which authors can make software available in a way that meets the principle of publication. Some authors explicitly place source code in the public domain with no restrictions, as they would materials with no commercial value. Others copyright and distribute their source code under an open-source license that grants some copying, redistribution, and modification rights while allowing other authors to build on the work. In a third mechanism, which minimally satisfies the principle of sharing publication-related information, an author copyrights the source code and provides it to a requestor at no cost but with no license to copy, redistribute, or modify the code.

A common current practice is to license published software to researchers at academic or other not-for-profit institutions for free or at minimum cost, while charging for-profit entities more of a "market rate" for access to the same published software. This practice is long-standing and workable. Many argue that it is a fair practice, because it provides a convenient mechanism for companies to contribute to the costs of software development and maintenance.

However, during its deliberations the Committee noted that, in those cases where a specific software implementation is integral to a paper or is itself the result announced by the paper, a different standard should apply. Consistency with the standards described herein for access to integral data and materials requires that such software implementations should be made available to the entire scientific community on the same terms. The principle is that publication involves equal responsibilities and benefits to not-for-profit and for-profit researchers alike.

In summary, this is an area where reasonable people disagree. The consensus view of the committee is that the software community should work toward providing equal access to software that is integral to a publication, while realizing that the practice of differential pricing is widely accepted.

In any case, differential software licensing terms are not problematic when a specific implementation is not integral to the publication. For instance, if a paper's result is an algorithm that is clearly and reproducibly described, then a software implementation of that algorithm might reasonably be kept proprietary. Indeed, charging for-profit entities for access to academically developed software tools is a traditional source of funding in bioinformatics, and many companies do think that paying for academic software is a reasonable way to contribute to software research.

Opinion at the workshop was deeply divided with regard to what constitutes reasonable access to data when a paper announces the existence of a database or a dataset too large to publish in print form. Some participants from academic and commercial institutions argued that when a dataset or database constitutes the main result of a paper, it should be made available on the same basis as though it were in the paper itself—broadly accessible at no cost, without restrictions, with no distinctions made between academic and industrial users of the data, and without a material transfer agreement or license.

Other participants said that insisting on such criteria would discourage some authors, particularly those in the for-profit sector, from publishing databases (or other large datasets) that they have compiled often at substantial cost and without direct public subsidy. They argued that the scientific community is better off gaining access to such information under restrictive conditions than not gaining access at all (Patrinos and Drell, 2002). According to that view, valuable information is being collected in the for-profit sector, and the research community must consider companies' need to protect the value of their property and should devise ways to promote dissemination of the information under peer-reviewed conditions. Mark Adams, of Celera Genomics, asserted it can be done through mechanisms "that are no more onerous than those already applied to materials." "It is very reasonable," he said, "to presume that there could be a database industry that publishes and for which subscription is deemed to be an entirely sufficient form of access."

It was argued, moreover, that data printed in journals are not truly free in that one must pay for a subscription to a journal or indirectly support a library that provides access to it. In that context, subscription to a database would be analogous to a journal subscription fee or commercial literature databases, such as ISI and Lexus-Nexus (which, however, are quite different than peer-reviewed scientific literature). However, most database fees are likely to be far more expensive than individual journal subscriptions; and once a journal subscription is paid for, it does not seem reasonable for additional fees to be imposed to gain access to data reported in it.

Given those views, several issues need to be sorted out. One is the question of whether publishing one's results automatically places them in the public domain. Can making data freely accessible mean "at no cost but with restrictions on use"? Could making data accessible mean "available, but not necessarily at no cost"? How should an author's commercial interests in the data be protected? And there is the underlying question of whether placing *any* restrictions on the use of data that are central or integral to a paper violates the *quid pro quo* that is at the heart of scientific publication—to provide access to information or materials essential to support and build on the major claims made in a paper in exchange for recorded recognition and acknowledgment of scientific accomplishment.

Because the cost of disseminating data on-line is negligible, it is reasonable to expect that data that are central or integral to a paper should be provided at no cost. However, making that data freely obtainable does not obligate an author to curate and update it. While the published data should remain freely accessible, an author might make available an improved, curated version of the database that is supported by user fees. Alternatively, a value-added database could be licensed commercially.

It is also important to reflect on the type of dataset or database that is put forward for publication and on whether there is an accepted method for pooling those data. In some fields of the life sciences, the research community has established public repositories to facilitate sharing of large datasets. By their nature, these repositories help to define consistent policies of data format and content and of accessibility for the scientific community. For example, standards for sharing published microarray data are in development, and biological taxonomists are promoting a central repository (MorphoBank) for morphological images. Structural biologists have agreed to deposit atomic coordinates of three-dimensional protein structures determined with x-ray crystallography or nuclear magnetic resonance spectrometry in the Protein Data Bank (*www.rcsb.org/pdb*). In genomics, the community standard is for researchers to deposit DNA sequences in one of the public electronic databases in the International Nucleotide Sequence Database Collaboration, which comprises GenBank, the European Molecular Biology Laboratory Nucleotide Sequence Database, and the DNA Data Bank of Japan (these are henceforth referred to collectively as genome databanks). The pooling of data in a common format is not only for the purpose of consistency and accessibility; it also allows investigators to synthesize new datasets and to gain novel insights that advance science.

If verification of data were the only concern, the data underlying a paper could be provided in "static" form and made available for viewing in a format chosen by the author. But for the research community to use and build on the results of a paper and to advance science (and its commercial applications), which is the ultimate purpose of today's system of scientific publishing, an increasing fraction of data must be available in "dynamic" form. That is, it must be possible to use the data in their entirety; to search, interrogate, rearrange, and manipulate the data; and to extract them from one program or framework and insert them into another.

The sequence data in public genome databanks are the starting point for an interconnected web of bioinformatics data resources that serve the larger research community. These resources include the National Center for Biotechnology Information BLAST server, a widely used resource that allows researchers to scan all known, publicly available DNA-sequence data for unexpected and informative homologies; public protein databases derived from the DNA-sequence data in GenBank, such as SWISS-PROT or Protein Information Resource; and public genome browsers, such as Ensembl, which adds useful annotations to eukaryotic genome-sequence data and facilitates their interpretation. Those and other public resources rely on free redistribution and creation of derivations of the underlying genome-sequence data. From that perspective, and as a matter of principle, it is important that scientific data related to a publication be made available in the public domain via an accepted repository identified by the community.

It is not difficult to recognize that some large datasets or databases will have commercial value. Some companies have identified a lack of protection for databases as the reason they are not willing to allow their researchers to publish on the same terms as other authors and the basis for requiring investigators who want publication-related data to sign an agreement about their use of the data. (See Box 3-1.) It is very much in the interest of the life-sciences community to foster solutions that increase access to scientific information, but the database legislation proposed in the Committee on the Judiciary of the U.S. Congress thus

BOX 3-1
Database Protection

Some developers of large data sets in the life sciences are reluctant to publish scientific findings related to their databases without the ability to prevent the data from being commercially exploited by others. But data themselves cannot be copyrighted, a principle reinforced by the 1991 Supreme Court decision in *Feist Publications, Inc. v. Rural Telephone Service Co.* (499 U.S. 340, 1991) which found that the underlying information in Rural's white pages telephone directory—that is, names, addresses, and phone numbers—are only factual information presented in an unoriginal arrangement. The Court ruled that, although Rural may have spent considerable time and expense in compiling the information, its database was not copyrightable and Feist (or anyone else) was free to copy it. Therefore, although the creative elements of databases—for example, the selection, coordination, and arrangement of the information—can be copyrighted, the facts themselves are ineligible for copyright protection.

When the substance of a database is eligible for copyright (because it is an original work of authorship), scientists can generally make use of a limited amount of that information because of a "fair use" exception that permits use of the material for such purposes as teaching, scholarship, and research. However, database owners—including companies such as Reed Elsevier, eBay, the National Association of Realtors, and Celera Genomics—are concerned that copyright does not afford enough legal protection to prevent their databases from being copied, modified, and sold commercially.

In 1998, the European Union's (EU) Directive on the Legal Protection of Databases (European Union, 1996) came into force, providing 15 years of protection for the contents of a database and each significant update and permitting database owners to prevent the use of substantial parts of it. The directive also has a reciprocity clause, which states that only countries that offer similar protections to EU nationals will receive this new level of protection within the European Economic Area.

Since 1996, several similar database-protection bills have been introduced by the Committee on the Judiciary in the U.S. Congress, but none has become law. Opponents of the bills—a loose coalition of scientific groups (including the AAAS and the National Academy of Sciences), universities (including the Association of

American Universities), libraries, telecommunication companies, Internet service providers, U.S. Chambers of Commerce, and value-added database producers—questioned the need for additional database protection, given the absence of documented cases of database piracy and the likely harm to science and education. If passed, the bills would have inhibited the free redistribution of data and derivatives of data. Database protection for government data would be available to database aggregators under some versions of proposed legislation, such as the bills of the Committee on Judiciary introduced in the 105th Congress (Coble, 1999). Large scientific datasets, once collected primarily by the federal government, are increasingly collected by for-profit concerns and made available under contracts or licenses that restrict the use of the data to approved individuals or for specific purposes. In some fields of science, contracts for data prohibit normal scientific practices, such as sharing the data with colleagues, publishing them in scientific journals, or using them to address more than one scientific problem.

In April 2001, legislative negotiations on database protection began anew in the House Committee on the Judiciary and Committee on Energy and Commerce. One draft bill would bar misappropriation of commercial data in which companies have made a substantial investment. Critics are concerned that this bill would give database owners almost complete control over the factual information in databases, which is not covered under copyright. In addition, critics are opposed to the bill's provisions for criminal penalties and large fines for misuse. In many scientific fields, the creation of derivative and integrative databases that combine data from multiple sources is a key part of scientific inquiry. A recent National Research Council study expressed concern that proposed legislation take into account the need to promote access to science and technology data and databases. The report noted that new federal legal protection against wholesale misappropriation of databases might be appropriate, but that any database protection adopted preserve the existing legal rights of "traditional and customary scientific, educational and research uses" of databases (NRC, 1999).

The life-sciences community should consider whether carefully crafted database protection might encourage the creation and publication of large datasets by affording database owners needed protections that do not impinge on the ability of the research community to communicate and share data, and that are consistent with the principles of publication.

far would have the opposite effect, inhibiting scientists' ability to use and distribute data and create derivative databases (NRC, 1999). The life-sciences community should help to ensure that any new database protections proposed are consistent with the principles of publication.

When companies have published papers in which a database was a central part of the research finding (and were granted an exception to the requirement to place the data in a public data repository), access to the data required an investigator to agree to terms that not only prohibited the use of the data for commercial purposes, but also prohibited other specific uses of the data (see Box 3-2), a fact that weakens the rationale

---

BOX 3-2
On Access to Published Genome Sequences

Under the terms of the public-access agreement that allows academic researchers to use Celera's human genome sequence (Venter et al., 2001) the data cannot be reproduced, redistributed, or used to prepare derivative works. In April 2002, the draft genome sequence of the *japonica* subspecies of rice was published in *Science* (Goff et al., 2002) under a similar agreement by a research team from the Torrey Mesa Research Institute (TMRI), a subsidiary of Syngenta, a Switzerland-based agricultural biotechnology company. In both cases, separate access agreements are required for academic and commercial researchers who wish to use the data.

The genome sequences from the Celera and the TMRI papers (except individual gene sequences that the companies have agreed to deposit in a genome databank, such as GenBank, if a journal requires it when a researcher wants to publish a paper about a specific gene) might not appear in a public genome databank or any other public bioinformatics resource and will not be available to the many researchers who use the National Center for Biotechnology Information BLAST server. In other words, the public-access agreements for Celera's human genome-sequence data and TMRI's rice genome-sequence data permit only "static" access, enabling verification of the paper's results. Depositing the data in a public genome databank would provide "dynamic" access and enable further research. The Celera and TMRI papers, therefore, are not consistent with the principles laid out in this report.

In contrast, in May 2002, Celera published in *Science* a comparative analysis of the human genome with its sequence of chromosome 16 of the mouse (Mural et al., 2002). The sequence of chromosome 16, generated as part of a shotgun assembly of the whole genome of the mouse, was deposited in the DNA Data Bank of Japan, the European Molecular Biology Laboratory Nucleotide Sequence Database, and GenBank, and thus made available to all, and additional information is provided on Celera's Web site. Access to Celera's whole-genome shotgun sequence of the mouse is available by subscription. This arrangement satisfies the core principles of freely sharing published data while allowing the company to commercialize related sequence information that is not central to the publication.

for making an exception in the first place. Considering that databases could be made available to different users under an array of terms outside the context of publication (one example is a subscription to Celera's Discovery System™), it is not altogether clear that compromising the *quid pro quo* will be offset by a gain in published research results that could not be made available by other means.

It may not be feasible to exert property rights in data that allow them to be published, verified by the scientific community, and provided in "dynamic" format without also facilitating commercial competitors. However, placing restrictions on the use of the data, charging an access fee, or making it difficult to compare with other datasets defeats the purpose of publication, because the data cannot be verified and the ability to build on it is diminished. These are factors that reviewers should consider when evaluating whether a submitted paper is important for the community.

As described in Chapter 2, however, companies do benefit from publishing, so it is not likely they will make all their data available only by subscription. It is also possible for a company to publish some data (without restricting access) that would increase interest in a more comprehensive database that is made available by subscription, as Celera has done (See Box 3-2, paragraph 3).

In its exploration of sharing publication-related data and software, the committee identified the following principles of publication:

> **Principle 1.** Authors should include in their publications the data, algorithms, or other information that are central or integral to the publications—whatever is necessary to support the major claims of the paper and to enable someone skilled in the art to verify or replicate and build on the paper's claims.

> **Principle 2.** If central or integral information cannot be included in a publication for practical reasons (for example, because a dataset is too large), it should be made freely (without restriction on its use

for research purposes and at no cost) and readily accessible through other means (for example, on-line). Moreover, when it is necessary to enable further research, central and integral information should be made available in a form that enables it to be manipulated, analyzed, and combined with other scientific data.

Principle 3. If publicly accessible repositories for data have been agreed on by a community of researchers and are in general use, the relevant data should be deposited in one of them by the time of publication.

As a way to improve the process of sharing publication-related data, the committee makes the following recommendation:

Recommendation 1. The scientific community should continue to be involved in crafting appropriate terms of any legislation that provides additional database protection.

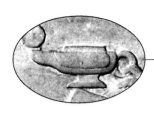

CHAPTER FOUR

Sharing Materials Integral to Published Findings

Sharing of materials integral to a published work is a responsibility of authorship. For consistency with the spirit of the uniform principle for sharing integral data and materials expeditiously (UPSIDE), materials described in a scientific paper should be shared in a way that permits other investigators to replicate the work described in the paper and to build on its findings. Sharing facilitates new scientific and commercial advances, eliminates duplicative efforts by others to recreate materials, and speeds the progress of science.

Making publication-related materials available begins logically with providing readers with information about them. For example, authors might anticipate which materials integral to their publications are likely to be requested and state in the "Materials and Methods" section or elsewhere how to obtain them. (It is appropriate, in this context, for scientific reviewers of papers submitted for publication to help to identify such materials.)

Some authors find that providing access to publication-related data and material is increasingly time-consuming and expensive. Creative solutions to solve these problems need to be found, and requestors of materials should accept some responsibility for the cost of distribution. A frequently requested reagent could be made reasonably available by an author's laboratory for a modest fee to cover the costs of production, quality control, and shipping. Repositories are another option. For example, plant germplasm can be deposited in distribution centers, cell lines could be submitted to the American Type Culture Collection, and mutant mice and the means to re-derive them could be sent to the Mouse Mutant Regional Resource Center or the Jackson Laboratory. In some fields, the use of repositories is a mandatory requirement of publication. Vouchers for studies in systematic and evolutionary biology, including paleontology, are deposited in museums and similar facilities, where they are available for further study and reinterpretation. Without the deposited vouchers, it would be impossible to verify and build on the results of papers in those fields.

Distribution centers are desirable, but not every reagent or biological material can be disseminated through them. Although the community pays a premium when an author uses a commercial licensee to distribute materials, as long as the materials are provided at a reasonable cost, this method of distribution is generally acceptable. Public granting institutions should be receptive to requests for supplemental funds to support the distribution of materials in high demand. An author has an obligation to anticipate requests and make arrangements for responding to them over time, but if this proves impossible, an author should provide requestors with information on how to reconstruct the material or on how the material was obtained.

When publication-related materials are requested of an author, it is understood that the author provides them (or has placed them in a repository) for the purpose of enabling further *research*. That is true whether the author of a paper and the requestor of the materials are from the academic, public, private not-for-profit, or commercial (for-profit) sector. Authors have a responsibility to make published materials available to all other investigators on similar, if not identical, terms. Prospective authors should be aware that publishing implies an obligation to share patented or copyrighted materials for research purposes. Therefore, if intellectual property rights are associated with a material, a research license should be made available to investigators who request it. If an author does not have rights to distribute the material, the contact information for its original source should be supplied.

The act of publishing information related to a patented material or process should not destroy the commercial interests of an author. However, an author (and an author's institution) must be willing to enable the community to use and build on that information by conducting further research. A good example is MIT's patented process for small interfering RNAs (siRNAs). Individual investigators can seek a research license for the process directly from MIT but can also obtain one by purchasing RNA oligonucleotides from vendors commercially licensed by MIT to incorporate a research license for MIT's process in the vendor's product. MIT is also considering whether to negotiate commercial licenses with companies that wish to use process in the development of new therapeutic products (as opposed to research reagents).

MATERIAL TRANSFER AGREEMENTS

Since the 1980s, transfers of materials between investigators (and between investigators' institutions) in the life sciences have been routinely accompanied by material transfer agreements (MTAs). An MTA defines the rights of the recipient to the use of a material, such as the right to undertake modifications, explore new uses of the material, and seek new inventions by using the material. It also defines how rights to intellectual property resulting from use of the material, if any, are apportioned between the material provider and the recipient.

Transfers of materials between parties are not always related to requests to authors after publication of a scientific paper. The terms of MTAs negotiated for transfers unrelated to publication may be, for various reasons, complex and far-reaching. Such MTAs may contain extreme restrictions on the use of a material, give the provider title to any new inventions that result from its use, include requirements to review the recipient's research progress related to use of the material, and have other terms. MTAs of this kind are often negotiated in the transfer of (unpublished) materials between companies and universities, but also occur in other contexts—for example, countries that grant permission to allow the export of biological samples for scientific study often prohibit their subsequent use for commercial purposes or distribution to commercial entities. Transfers of that type between two institutions outside the publication process were not discussed at the workshop and are beyond the scope of this report. They stand in contrast with MTAs that are negotiated as a direct consequence of a request for material related to a published work.

If the transfer of a material described in a scientific publication requires an MTA, an author should provide, in the appropriate section of the paper, the URL of a Web site where the MTA can be viewed. According to the principle of publication, an author has an obligation to ensure that material will be provided to other investigators under terms that do not block their ability to engage in the research necessary to replicate the author's work or build on the publication's findings. MTAs can support the objectives of publication and sharing by incorporating a minimal set of straightforward terms that do not interfere with the conduct of research. Terms that are consistent with the spirit of sharing publication-related materials and that acknowledge the provider's contribution of a material include

- A requirement that the provider be acknowledged as a source of the material in any publication by the recipient.
- A prohibition against use of the material for work with human subjects, including diagnostic testing, if the material has not been approved for use with human subjects.
- An acknowledgment that transfer of the material does not affect the legal title to it.
- A requirement that the recipient not disseminate the material to others outside his or her laboratory without the provider's permission.
- A requirement that the material be used only for research purposes; that is, where the primary intention of the research is the fundamental increase in knowledge. This excludes its manufacture for sale, licensing to others, or contract research undertaken for a commercial concern.
- A general description of the research for which the recipient

intends to use the material (to eliminate frivolous or irresponsible requests).

- A provision that disclaims any warranties on the material and excludes the provider from any liability for damages that arise from the use, storage, transport, or disposal of the material by the recipient, including liability related to the recipient's infringement of any third-party intellectual property rights.

Terms that burden the recipient with unnecessary requirements or impede the transfer of materials are inappropriate in an MTA associated with publication-related materials and defeat the intention of sharing them. Unacceptable terms include

- Requirements for periodic reporting of research findings related to the use of the material.
- Ownership by the provider of the recipient's data or other research results.
- Limitations on the recipient's ability to publish research results.
- Limitations on the recipient's ability to discuss research results with his or her laboratory members before publication.
- Automatic coauthorship rights for the provider.

Terms like those slow the progress of science by unfairly extending the provider's rights beyond the material into the recipient's work and by interfering with a recipient's participation in research-related activities. For example, a recipient of materials may through their own research arrive at conclusions that differ from those published by the provider of the materials; if required to collaborate or coauthor downstream publications, they would be inhibited from communicating their contrary findings, much to the detriment of the scientific process. Requiring prepublication review—a common term in MTAs currently used by many institutions—is inappropriate if the transfer involves only materials related to a published paper, as opposed to a situation in which additional confidential information about the material will be provided to the recipient. It is a courtesy to send the provider of a material a copy of a paper before publication, but it should not be a requirement of the MTA. Similarly, it is a courtesy, but should not be a condition of sharing, to require the recipient to notify the provider if patent applications will be filed on new inventions related to the material or its use.

RIGHTS TO NEW INVENTIONS

MTAs that accompany transfers of a publication-related material that might have commercial potential raise more complex issues. Those MTAs will inevitably focus on the provider's rights to the potential fruits of research realized through the recipient's use of the material. These include what are called new inventions: *improvements* of the material, *new uses* of the material, and *new substances* created through the use of the material. Specific terms related to those types of new inventions are often a source of contention during the negotiation of MTAs.

The UPSIDE principle can be used to resolve some issues raised in MTAs by identifying clearly unacceptable terms that intrinsically block the recipient of a material from doing research to replicate and build on the author's findings. For example, it is not acceptable for an author to demand an *exclusive* license to commercialize a new substance that a recipient makes with the material. A recipient's ability to conduct research might require the use of multiple materials from different providers, and as a matter of logic, the recipient cannot grant multiple exclusive commercial licenses to providers of all the materials used in making the new substance. Therefore, a provider's demand for an exclusive license to a new substance effectively blocks the recipient from assembling the materials needed to build the edifice of science.

There is currently no community standard that addresses a provider's request for a commercial license to the recipient's improvements to or new uses of a publication-related material. Transfers between academic institutions of materials created with National Institutes of Health (NIH) funding do not generally include *any* provisions granting the provider a license to commercialize inventions made with transferred materials. NIH has put forward a uniform biological MTA and simple letter agreement as templates for transfers of materials arising from NIH-funded research (NIH, 1999). Although not necessarily appropriate for MTAs negotiated outside the context of NIH-funded research, those models are considered to be reasonable vehicles for the transfer of most published materials.

The negotiation of an MTA in which the provider seeks rights to new inventions made through the use of the material is the most challenging. The lack of standardized terms for dealing with improvements and new uses of publication-related materials is a reflection of the commercial interests of the institutions involved in the transfer, whether they are for-profit or not-for-profit entities. Those interests overwhelm the application of the principle of publication as a means to guide the negotiation of MTAs.

Therefore, other factors have evolved as the basis of what an author can acceptably request in an MTA. For example, when a provider's material is protected by patents, a request for an exclusive commercial license to improvements and new uses of the material is often deemed acceptable because the provider can often legally block others from exploiting those new inventions anyway. If the material is not patented, requests by the provider of a material for exclusive commercial licenses to improvements and new uses made by the recipient may be accepted or rejected by a recipient's institution, depending on whether the provider plans to contribute in some way to the recipient's work with the material. If the provider will not contribute funds or other types of support, such a request is often rejected because it is viewed as an overreaching claim for reward for work neither done nor envisaged. If the provider of a material requests a *nonexclusive* commercial license to inventions made through the use of the material, negotiations between institutions often focus on the specific facts of the case, such as the true importance of the material provided, how integral it will be to the recipient's prospective work and whether the new inventions might block the provider of material from exploiting the original material. Some negotiations may elaborate whether, in exchange for the right to a commercial license, the provider must pay royalties to the recipient.

There are many contrasting views in the technology-transfer community about what is appropriate in an MTA in any given situation and for any particular material. Because there are so many nuances in the negotiation of these issues (the examples given above are not exhaustive), there is a potential for delay in reaching agreement, and sometimes there is an impasse. The proliferation of MTAs with idiosyncratic requirements set by multiple institutions is, in the end, an impediment to sharing publication-related materials.

The purpose of sharing publication-related materials is to enable research—that is, to allow the recipients of material to replicate and build on the work of the authors—and the terms of MTAs and their negotiation should not create a barrier to this goal. All institutions engaged in technology transfer should closely examine the merits of adopting a standard MTA, and efforts to streamline the process need to be championed at the highest levels of universities, private research centers, and commercial enterprises. The NIH guidelines for sharing research resources are useful as a starting point for the discussion (NIH, 1999).

In the meantime, both practicality and expediency should be embraced in procedures for transferring publication-related materials. Commercial and proprietary interests notwithstanding, an author's fulfillment of the *quid pro quo* includes ensuring that MTAs related to materials described in the author's publications are not burdened by protracted negotiations over terms. As a best practice, a period of 60 days for the negotiation of publication-related MTAs would eliminate uncertainty for requesters of materials and remove what is currently perceived as a substantial barrier to the ability of investigators to move forward with their research plans. Institutions should consider whether MTAs are necessary for every material transferred, and universities, in particular, should provide adequate resources to support the effective and expeditious management of MTAs. If sharing publication-related materials in a timely fashion is important to participants in the publication process, participants should encourage their institutions to commit to achieving that goal.

In its exploration of sharing publication-related materials, the committee identified the following principles of publication:

> **Principle 4.** Authors of scientific publications should anticipate which materials integral to their publications are likely to be requested and should state in the "Materials and Methods" section or elsewhere how to obtain them. If an MTA is required, the URL of a Web site where the MTA can be viewed should be provided. If the authors do not have rights to distribute the material, they should supply contact information for their original source. A frequently requested reagent can be made reasonably available in the commercial market or by an author's laboratory for a modest fee to cover the costs of production, quality control, updating, and shipping.

> **Principle 5.** If a material integral to a publication is patented, the provider of the material should make the material available under a license for research use.

The following recommendations address actions that, if adopted by the life-sciences community, would expedite the process of sharing publication-related materials:

> **Recommendation 2.** It is appropriate for scientific reviewers of a paper submitted for publication to help to identify materials that are integral to the publication and likely to be requested by others and to point out cases in which authors need to provide additional information on obtaining them.

> **Recommendation 3.** It is not acceptable for the provider of a publication-related material to demand an exclusive license to commercialize a new substance that a recipient makes with the

provider's material or to require collaboration or coauthorship of future publications.

Recommendation 4. The merits of adopting a standard MTA should be examined closely by all institutions engaged in technology transfer, and efforts to streamline the process should be championed at the highest levels of universities, private research centers, and commercial enterprises.

Recommendation 5. As a best practice, participants in the publication process should commit to a limit of 60 days to complete the negotiation of publication-related MTAs and transmit the requested materials or data.

CHAPTER FIVE

Different Interpretations of Existing Standards

A fundamental issue in the debate over the sharing of publication-related data, information, and materials is whether exceptions should be made to a community standard if the progress of science might be advanced by them. Participants at the workshop expressed different opinions as to whether a narrow or broad interpretation of a putative community standard is appropriate. To illustrate more clearly the strain that exceptions place on the publication process, some of the common arguments in favor of exceptions are presented here and examined in the context of the principles of publication put forward in this report.

The essential "value" of a scientific publication resides in the scientific finding and its implications as seen by the peer scientific community. If the finding is viewed as important, a journal's readers may be willing to accept partial disclosure of supporting data and materials by the author so as not to delay awareness of the finding having been made.

Journal editors make decisions routinely about how much information the authors must provide to support their findings—in part on the basis of what will satisfy their readers. But progress in science is most efficient when an author's peers can critically evaluate published findings. Announcing a finding without making available the data or materials that are integral to the publication (that is, necessary to support the major claims of a paper and allow knowledgeable peers to validate or refute the major claims) may be appropriate for an advertisement or press release, but it is not appropriate for scientific publication.

An author should only need to disclose or share that which is required to reproduce and validate the published result, nothing more, nothing less.

Presenting results with enough detail so that they can be repeated might be the minimum a journal officially requires of an author but it is substandard from the perspective of the community, in particular, if repeating the work will be labor intensive. Scientists do not read others' papers in order that they might repeat those experiments; rather they read articles to find insights and gain knowledge that allows them to move forward from that point. Only when scientists are unable to successfully build on results of a paper are they inclined to repeat the author's experiments. Taking the stance that authors need only furnish what is necessary to repeat one's experiments removes the value of the cumulative process of science and, considering how science is conducted today, is unrealistic. It is not possible, for example, to get a public research grant to repeat the experiment of another scientist.

Partial access to data in a publication is better than no access at all.

It has been proposed that providing data on a private Web site, with no limit on what can be viewed but with limits on the amount of data that can be downloaded at one time, satisfies the *quid pro quo* of publication. However, if the data are central or integral to the reported findings, this arrangement violates the spirit of the fundamental principle of allowing other researchers to replicate, verify, and build on the findings. Researchers need to be able to manipulate, query, and transform the data that support a publication's findings so that they can build on them.

Some categories of materials are difficult, time-consuming, or expensive to reproduce; therefore, requiring authors to share them is unreasonable.

The community has never required an author to provide extensive or ongoing technical support for a requester of materials. If materials are scarce or difficult to replicate, a request could be reasonably met by providing requestors with detailed protocols and advice regarding their synthesis, or with information on how they were obtained. Stock centers and repositories (such as museum collections and herbaria, the Jackson Laboratories and the American Type Culture Collection) constitute another appropriate alternative for distribution or maintenance of certain materials and are standard for systematic and evolutionary biology, including paleontology.

Because large-scale data assemblies can be extremely expensive to produce initially, authors should not have to make them available free.

Part of the responsibility of publishing is to share what is integral or central to the findings of a publication in exchange for credit and acknowledgment of research achievements. The commercial market has mechanisms other than publishing for making data available by subscription. The costs associated with distributing and updating a publication-related dataset might reasonably be charged to users if no public archive is available, but it is not appropriate to impose a charge to recoup the original costs of production. In the future, National Institutes of Health (NIH)-funded scientists might be able to request supplemental support to an existing grant to help distribute data (NIH, 2003).

It is not always possible to distribute materials or freely share data because of prior contractual agreements made with a research sponsor.

If a contractual agreement (for example, between an academic researcher and an industrial sponsor of the research) prohibits such sharing, the researcher should not publish. Researchers who wish to publish should avoid entering into contractual agreements that prevent unconstrained sharing of publication-related data or materials.

Researchers should have an exclusive right to analyze, or "mine," the data they produce for a specified period after publication. The delayed release of full datasets in some disciplines should be permitted.

Science moves forward most rapidly when the research community has the ability to view and use all the data integral to a published research finding immediately on publication. There is precedent for placing a time-limited hold on some aspects of data, such as atomic coordinates in crystallography, but the discipline has moved steadily away from that practice. The adoption of such a moratorium by a particular community can at best be justified only as a temporary interim step toward the goal of full release upon publication.

Young investigators will easily be scooped and their careers potentially will suffer if they are required to share data or materials related to their publications. Researchers just starting their academic careers should be granted a moratorium on sharing to prevent them from being overrun by their competitors.

Participating in the scientific enterprise involves agreeing to "do the right thing," which entails some risks. Young investigators abide by community standards and risk facilitating their competitors' research because there is an equal, if not greater, probability that they will be the beneficiaries of unrestricted sharing by others. A principal investigator can certainly try to protect the interests of graduate students or postdoctoral fellows, for example, by asking other investigators to become collaborators, and to use data or materials other than those being used by a young investigator in the laboratory. However, the sharing of materials cannot be made contingent on a promise by the recipient to enter into collaboration or to avoid competing with the one who supplies the materials. Furthermore, granting a special exception to some researchers is problematic for a variety of reasons, including the difficulty of deciding who qualifies for the exception. Who is more vulnerable: a starting researcher who has just finished a postdoctoral position in a famous laboratory, or a late-career researcher whose only research grant has just been turned down for renewal? The exposure to both benefit and risk associated with competitive activities triggered by publication must be shared equally by all participants in the publication process.

Authors should have the right to request a collaboration or coauthorship of future publications in exchange for publication-

related materials, particularly if the materials are scarce or difficult to produce.

One of the advantages of publishing is that it may lead to new and fruitful collaboration. Authors can pursue mutual research interests with requestors of materials, but it is not reasonable for an author to demand a scientific collaboration or coauthorship of a future publication in exchange for material. The principle of publication implies that authors must make their materials accessible on terms that do not interfere in a recipient's work. However, authors who provide data or materials should be acknowledged for their contributions (see Chapter 6).

The life-sciences community must recognize the growing role of the for-profit sector in basic research and acknowledge differences in culture and tradition between academe and industry. If the for-profit sector is not treated exceptionally, it may choose not to publish and instead charge for access to its data or materials. Moreover, some fields of the life sciences, such as plant biology, have received less public funding than others; in these fields, the for-profit sector is likely to be generating the most cutting-edge data and scientific findings. If there is no competing public effort that will make the same data public eventually, it is not realistic to think that companies will disclose data without some incentive or special exception.

By requiring sharing of publication-related materials and data, the scientific community does risk the possibilities that the for-profit sector will choose not to publish some of its findings and that some data or materials will be made available only at a cost. But allowing data to be provided on terms that don't meet community standards also has a cost—the accumulation of fragmented data sets that are difficult to validate, compare, search, or combine with the data in public repositories. In addition, allowing data to be made available to companies on different terms from those given to academics places another burden on the publication system. That fact might not concern the academic community, but it is a double standard that makes it difficult to expect the commercial sector to share data and materials as "freely" as academics claim to do.

There may also be data and materials in the for-profit sector of which the larger community is not even aware, because it is too commercially valuable to be shared. It is possible, therefore, that loosening standards for sharing publication-related data and materials (for example by allowing companies to make them available only under restrictive licenses) will not lead the for-profit sector to disclose via publication much more than it does now, because such disclosure is likely still to entail too many commercial risks.

Whether or not the short-term gain of partial and restricted access to data is worth the long-term setbacks to the system of publication is a matter of debate and difficult to prove. However, other ways exist to facilitate access to data or materials generated in the for-profit sector that might result in greater benefits to the scientific community than making exceptions to community standards and principles. These include, for example, the creation of private consortia (such as the SNP [single nucleotide polymorphism] Consortium) and public-private consortia (such as the consortium to accelerate sequencing of the mouse genome).

Requestors may use data or materials provided to them to compete, gain commercial advantage, or find a flaw in the original study and disprove its findings.

Competition and correction of erroneous conclusions and the later genesis of data that represent commercializable inventions are part and parcel of the scientific enterprise and part of the risk of publishing. Indeed, they are vital. Identifying problems in a flawed study improves the overall scientific process. Denying potential competitors access to data undermines the basic principles of sharing.

Authors should have the right to share publication-related data or materials with academic investigators only.

This view must be rejected as an artificial taxonomy. There should be a single scientific community that operates under a single set of principles regarding the pursuit of knowledge, a common ethic with regard to the integrity of the scientific process, and a long-held commitment to the validation of concepts by experimentation and later verification or falsification of published observations. There is no clear line between "for-profit sector" and "academic" research. Some of the research done in companies is basic research with no predictable commercial end point, whereas some academic laboratories are directly connected to companies via sponsored research agreements, collaborations, consulting agreements, or stock ownership.

In the systematic and evolutionary biological community, nations often restrict the right of commercial firms to investigate or use biological samples for which they have allowed export. Because it is a matter of law, and the material is unequivocally the property of the nations involved, investigators are obligated to abide by these restrictions. While there is interest in pursuing the use of such materials by commercial firms in a regulated legal context, at the present time, companies are compelled to negotiate individual separate agreements with a country to use biological samples of interest.

Aside from this unique situation, the committee became convinced during its deliberations that exceptions to standards in one form or another could not be rationalized without sacrificing the integrity of the principle of publication. In considering arguments for making exceptions to community standards, including the need to accommodate commercial interests, the costs of producing data and materials, the vulnerability of young investigators to competition, and an investigator's right to mine his or her data before others, the committee found that participants in the publication system were just as likely to benefit as to be hurt by a system that favored the sharing of data and materials. In some instances, avenues other than publication are available for those investigators who want to publicize their findings while maintaining control of the related data. In other cases, reasonable and innovative ways can be found to overcome the problems of costs, contractual restrictions, and competition. Notwithstanding the rule of law and other common-sense situations, exceptions unfairly penalize the community, which would have otherwise had access to the data, information, or material being withheld. Furthermore, granting a special exception to certain categories or particular researchers is problematic for a variety of reasons, including the difficulty of deciding who qualifies for the exception. Considering that publication standards maintain quality and facilitate the work of the community in moving science forward, the committee observed that exceptions are likely to weaken the effectiveness of that process over the long term:

Universal adherence, without exception, to a principle of full disclosure and unrestricted access to data and materials will promote cooperation and prevent divisiveness in the scientific community, maintain the value and prestige of publication, and promote the progress of science.

CHAPTER SIX

Encouraging Compliance with and Continuing the Development of Standards

COMPLIANCE AND ENFORCEMENT

When the principles of publication are not upheld, the scientific enterprise suffers. For example, a recent survey of academic geneticists reported numerous adverse consequences of their colleagues' withholding publication-related information, data, or materials (Campbell et al., 2002). Among the consequences cited were the inability to replicate published research, substantial publication delays, and abandonment of promising lines of research. In addition, a majority of geneticists felt that withholding detracted from communication in science, slowed scientific progress, harmed peer relationships, and adversely affected the education of students and postdoctoral fellows. Some reported that they had ended collaborations as a result of a colleague's withholding materials and data.

Other adverse effects of noncompliance with sharing requests were noted in discussions at the workshop. Researchers whose requests for publication-related materials or data are not granted may be unable to do the best science and thus may not remain competitive in their field. In addition, they may spend time and effort in unnecessarily repeating work done by others. Ultimately, some workshop participants felt that if problems of noncompliance are not resolved, the result will be a scientific culture that is rife with conflict.

The obligation for fulfilling community standards for sharing publication-related data and materials lies first with the authors of a publication. As a practical matter, the author designated in the publication as the corresponding author should be responsible for identifying which coauthor has the materials and other information requested by a third party and should confirm that they are provided when requests are made.

Workshop participants expressed a variety of views of how the scientific community can encourage authors to comply with their obligation to share. Several panelists emphasized that researchers should first take an informal, one-on-one approach to resolve compliance issues, because an initial lack of a response of an author to a request does not necessarily indicate ill intent. Requestors should first consider simple measures, such as telephoning an author to determine why he or she did not respond to a request, and seek resolution before expanding a dispute to involve other parties.

Institutions involved in the scientific enterprise—including journal editorial offices, universities, and funding organizations—should also assume some responsibility for ensuring that authors make available the resources that will enable other researchers to replicate, verify or refute, and build on reported results. If it becomes necessary for a requestor to move beyond straightforward overtures to a paper's authors, most of the workshop participants agreed that journals should assume primary responsibility for enforcement.

It is not known how many instances of noncompliance are ever brought to the attention of journal editors or other external authorities. However, with respect to cases that are reported to journals, journal editors at the workshop reported a high rate of success in getting authors to share published materials and data, noting that a telephone call or letter from the editor-in-chief or managing editor to an author is often sufficient to resolve problems. Many journal editors stated their willingness to enforce standards of sharing, but one editor expressed concern about adjudicating complicated disputes over the sharing of data and materials, particularly those involving legal wrangling over intellectual property issues.

Sometimes an author might have the responsibility *not* to honor a request for published materials or data—for example, if bioterrorism is suspected—until the motivations and credentials of the requestor are validated. Although that subject has not been well explored, it seems appropriate to have journal editors mediate disagreements in such situations.

In addition to helping to resolve cases of noncompliance after publication, scientific journals can play an important role in encouraging authors to comply with standards and the principles of publication *before* a paper is published by incorporating transparent standards into their official policies. As noted in Chapter 2, a recent examination of the instructions for authors of 56 life-sciences and clinical-medicine journals showed that the specifics of these policies vary considerably (Table 2-1). Furthermore, 25 of the 56 journals do not have any stated policy on sharing of data or materials in their instructions for authors, and clinical journals are more likely than life-sciences journals to lack such a policy.

Another factor that may contribute to noncompliance is that few journals—even among journals that have an official policy for sharing materials or data—provide any statement or policy guidelines as to the consequences for authors who do not comply. Furthermore, most journals do not provide a procedure for registering or publicizing complaints about noncompliance. Of journals that specify sanctions, most say that they would consider denying a noncomplying author further rights to publish in their journals; several participants considered this insufficient to engender compliance. Some workshop participants noted that peer pressure and opinion can be influential in bringing about compliance. A journal might choose to publicly declare an author's noncompliance (after all honest attempts are exhausted) in a specific section dedicated to this purpose.

Nicholas Cozzarelli, editor-in-chief of the *Proceedings of the National Academy of Sciences* (PNAS), provided commentary at the workshop about an "escalator policy" for handling complaints about noncompliance. If an author does not provide requested material after receiving a letter from PNAS, it will threaten not to publish future papers by that author. Finally, the PNAS will threaten not to let that author's coauthors publish again in the journal. Further workshop comment by Laurie

Goodman, executive editor of *Genome Research*, indicated that the journal would remove a paper from its on-line version and insert a note that the paper had been removed if an author failed to comply with the journal's policy for sharing materials or data.

Workshop participants discussed the role of funding organizations, universities, and other research institutions in enforcing compliance with standards of publication. According to Philip Campbell, editor-in-chief of *Nature*, "editors at the moment are acting in isolation in trying to impose sanctions." He said that knowing what sanctions would be vigorously supported if he contacted a noncomplying author's research institution or funder might make it easier to enforce compliance.

It is unclear to most scientists what procedures and policies universities have in place for ensuring that their investigators and staff comply with the community's expectations of authors. It is difficult to find current published policies laid out by universities or organizations that provide research funding regarding formal procedures for resolving problems of noncompliance by their employees or grantees. Although a telephone call or letter to an author from a program director or other representative of an organization can be effective in achieving compliance, funding organizations and universities, like journals, can encourage compliance earlier in the process by developing and enforcing transparent policies that encourage sharing of research resources. For example, the National Institutes of Health (NIH), National Science Foundation (NSF), and other funding organizations, such as the Howard Hughes Medical Institute (HHMI), have policies that reinforce and in some cases extend the standards set by the research community for depositing data in public databases. NIH has also issued a set of principles and guidelines on obtaining and disseminating biomedical research resources (NIH, 1999), although these are not tied specifically to publication. In addition, beginning in October 2003, researchers who receive funding from NIH will have to meet broad standards for data sharing. According to a statement issued by NIH in February 2003, "The NIH expects and supports the timely release and sharing of final research data from NIH-supported studies for use by other researchers. Consequently, investigators submitting an NIH application seeking $500,000 or more . . . are expected to include a plan for data sharing or state why data sharing is not possible" (NIH, 2003). To facilitate such sharing, NIH will allow researchers to request, in their original grant application or as a supplement to an existing grant, funds for sharing or archiving data.

One reason that researchers have cited for not sharing published materials is the time, effort, and cost involved in doing so. In some cases, that is a legitimate concern. NIH policy allows grantee organizations and investigators to charge requestors for the reasonable cost of production of biologic materials and for packaging and shipping, although income from these charges must be recorded as program income.

Other means exist to facilitate and minimize the costs of sharing publication-related research resources, including the deposition of research materials in existing public repositories, such as the American Type Culture Collection, and establishment of new repositories to facilitate sharing. NIH has established repositories to meet the needs of some specific research communities, such as the NIH AIDS Research and Reference Reagent Program (*www.aidsreagent.org*) and the Cancer Genome Anatomy Project (*cgap.nci.nih.gov*) And some researchers have established their own "cottage industries" for producing and distributing commonly requested materials.

A number of workshop participants advocated encouraging compliance with community standards by rewarding authors for sharing rather than by focusing on penalties for not sharing. They noted that whereas publication provides benefits for an author, the benefits of sharing publication-related resources are not always clear. For example, authors often fail to acknowledge those who have provided materials, data, or other information that helped in obtaining the findings they are publishing. Sharing should be recognized by citing a relevant publication of the donor of the material, and in the acknowledgement section of a paper. Another idea is to create a public database for acknowledgments. Such approaches would make it easier to recognize and reward those researchers who have been generous in sharing publication-related materials, data, software, or other information.

STANDARDS DEVELOPMENT

The existence of technical standards that allow data, materials, and other scientific information to be easily shared is critically important to the research communities that use those resources. Because they are so important, scientists organize circles to discuss standards in national and international workshops, at scientific-society meetings, and on electronic bulletin boards. For example, the Microarray Gene Expression Data Group calls itself a "grass-roots movement" to develop standards for DNA-array experiments and data representation so that investigators can compare and validate results from different arrays. Those kinds of efforts might be expanded to develop consensus on what data should be included in a publication and how data, materials, and information related to the publication should be shared, considering the needs of investigators in the particular field. Professional scientific societies, many of whom publish journals, also provide a natural forum for the discussion of community standards for sharing publication-related materials and data. Societies can help to identify repositories for specific types of data and materials that should be used by community members to facilitate sharing in different disciplines.

Crystallography: A Case Study

The development of standards for sharing publication-related data in crystallography provides a useful illustration of how one community of researchers has worked to establish norms that maximize the scientific value of information produced by individual investigators. It also shows how the research community, in concert with journals and funding organizations, continues to adjust standards and seek their enforcement as the field evolves.

In 1971, a group of crystallographers established the Protein Data Bank (PDB) as a worldwide archive for three-dimensional structure data on biologic macromolecules. At first, only a handful of structures were deposited in the archive each year. However, deposition of structures in

PDB began to rise dramatically in the 1980s because of technological advances in crystallography and changes in community views on data sharing. In 1989, the International Union of Crystallography (IUCr), which represents the crystallography research community, adopted a resolution calling on crystallographers to deposit atomic coordinates and related data in the appropriate structural database where a research article drawing conclusions from the data was submitted for publication. Because substantial time and effort are needed to solve a protein structure, the IUCr policy allowed authors to request a 1-year hold before public release of the data by the database so that authors could reap the intellectual benefits of their efforts.

By the early 1990s, as a result of the efforts of members of the crystallography community, many journals, NSF, and NIH had adopted the IUCr policy, including the 1-year hold allowance. The decision by HHMI to require deposition of coordinates by all its investigators was one of the critical steps in making the standards universal. Adoption of the standards by all key funding organizations and the most prestigious journals was pivotal in encouraging compliance.

In recent years, the time needed to determine the structure of most proteins has decreased dramatically, and standards for deposition and public release of structural data have continued to evolve. Interest in gaining access to macromolecular-structure data has grown as a result of the continuing development of new methods for analyzing the data and new experimental uses for them, including studies on protein folding, protein family organization, structure prediction, and drug design. In response to that interest, the PDB is evolving from a simple repository of data to one that provides mechanisms for researchers to understand biological function through investigation of sequence and molecular structure. As part of this evolution, the PDB is engaged in effort to achieve uniformity of format among all the records deposited in the archive since its inception.

As a result of the changes in the field, and with the vigorous support of scientists in the crystallography community, in the late 1990s some scientific journals and several special NIH research programs adopted an immediate-release policy for macromolecular-structure data. In January 1999, NIH announced a new policy requiring its grant recipients to arrange for immediate release on publication of data deposited in the PDB. Community standards for sharing these data are still in flux. Although a number of journals now require immediate release of structure data, some still allow authors to request a 1-year hold. Others have adopted the latest IUCr recommendations, which urge authors to release their data immediately after the publication date but still give them the leeway to request a delay of up to 6 months. In addition, standards and policies still vary with regard to the kinds of data from structure studies, other than atomic coordinates, that should be deposited in PDB.

There are analogies between the evolution of community standards in the crystallography community and the current debate about sharing genome-sequence data on publication. When NIH and some journals first considered requiring immediate release of protein-structure data, various arguments were made against changing the standard. Opponents of the change argued that authors needed the protection conferred by a 1-year delay and that companies might defer publication without such protection. They also cited the need to protect the interests of graduate students and postdoctoral fellows. In addition, some argued that the new standard would fail because journals would not be able to enforce it. Despite those arguments, several journals did adopt the standard and are enforcing it. Community standards in the field will no doubt continue to evolve.

Community standards are not federal regulations; rather, they are self-imposed by the community and sometimes incorporated in journal policies. The responsibility for developing and updating community standards lies with all members of the community who participate in the publication process and have an interest in the progress of science—academic, government, and industrial scientists; publishers and editors of scientific journals; and institutions and organizations that conduct and fund scientific research.

In many fields of the life sciences, uniform standards for reporting data are still being developed. However, it is the committee's view that once such standards have been established, journals should enforce them. Community standards, like the principles articulated in this report, are really only valuable to the extent that they are upheld by the scientific journals and honored by the community. The data generated by modern science may be increasingly diverse and complex and present novel challenges, but the power of the principles first established by Henry Oldenburg and the *Philosophical Transactions of the Royal Society* in 1665 remain undiminished: The rewards of publication counterbalance inclinations to secrecy. Oldenburg's simple idea created an ethic of open disclosure of scientific results that has lasted for centuries and served to move science forward. We hope this report, which reaffirms that ethic, will be a useful contribution to the community's discussions of standards for sharing publication-related data and materials.

With this philosophy in mind, the committee puts forward several recommendations for consideration and discussion by the community:

Recommendation 6. Scientific journals should clearly and prominently state (in their instructions for authors and on their Web sites) their policies for distribution of publication-related materials, data, and other information. Policies for sharing materials should include requirements for depositing materials in an appropriate repository. Policies for data sharing should include requirements for deposition of complex datasets in appropriate databases and for the sharing of software and algorithms integral to the findings being reported. The policies should also clearly state the consequences for authors who do not adhere to the policies and the procedure for registering complaints about noncompliance.

Recommendation 7. Sponsors of research should clearly and prominently state their policies for distribution of publication-related materials and data by their grant or contract recipients or employees.

Recommendation 8. If an author does not comply with a request for data or materials in a reasonable time period (60 days), and the requestor has contacted the author to determine if extenuating circumstances (travel, sabbatical, or other reasons) may have caused the delay, it is acceptable for the requestor to contact the journal in which the paper was published. If that course of action is not successful in due course (another 30 days), the requestor may reasonably contact the author's university or other institution or the funder of the research in question for assistance. Those entities should have a policy and process in place for responding to such requests for assistance in obtaining publication-related data or materials.

Recommendation 9. Funding organizations should provide the recipients of research grants and contracts with the financial resources needed to support dissemination of publication-related data and materials.

Recommendation 10. Authors who have received data or materials from other investigators that have contributed to the work published should appropriately and publicly acknowledge such contributions.

References

Campbell, E.G., B.R. Clarridge, M. Gokhale, L. Birenbaum, S. Hilgartner, N.A. Holtzman, and D. Blumenthal. 2002. Data withholding in academic genetics: Evidence from a national survey. *JAMA* 287(4): 473-480.

Coble, Hon. U.S. Representative. 1999. H.R. 354. Collections of information antipiracy act. 105th Congress.

Department of Health and Human Services. 2001. Federal Policy for the Protection of Human Subjects. 45CFR46.101-409.

European Union. 1996. Directive 96/9 of the European Parliament and of the Council of 11 March 1996 on the Legal Protection of Databases, O.J. (L77) 20.

Goff, S.A., et al. 2002. A draft sequence of the rice genome (*Oryza sativa* L. ssp. *japonica*). *Science* 296: 92-100, April 5, 2002.

Kennedy, D. and B.R. Jasny. 2001. (Editorial) *Science* 291:115,3 February 16, 2001.

Mural, R.J., et. al. 2002. A comparison of whole-genome shotgun-derived mouse chromosome 16 and the human genome. *Science* 296:1661-1671, May 31, 2002.

National Institutes of Health. 1999. Principles and guidelines for recipients of NIH Research Grants and Contracts on obtaining and disseminating biomedical resources. 64FR72090 (1999).

National Institutes of Health. 2003. Final NIH statement on sharing research data. NOT-OD-03-032. Release date: February 26, 2003.

National Research Council. 1992. Responsible Science: Ensuring the Integrity of the Research Process. Washington D.C.: National Academy Press.

National Research Council. 1999. A Question of Balance: Private Rights and the Public Interest in Scientific and Technical Databases. Washington D.C.: National Academy Press.

Patrinos, A. and D. Drell. 2002. The times they are a-changin'. *Nature* 417:589-590 (Commentary), June 6, 2002.

PL 96-517. 1980. The Bayh-Dole Act of 1980.

Venter, J.C., et al. 2001. The sequence of the human genome. *Science* 291: 1304-1351, February 16, 2001.

APPENDIX A

Committee Biographies

Thomas R. Cech, Chair
Dr. Cech is President of the Howard Hughes Medical Institute. He is also a Distinguished Professor at the University of Colorado, Boulder. He received his B.A. degree in chemistry from Grinnell College and his Ph.D. degree in chemistry from the University of California, Berkeley. His postdoctoral work in biology was conducted at the Massachusetts Institute of Technology. He is a member of the National Academy of Sciences and of the Institute of Medicine. Among the many honors he has received are the Lasker Award, the National Medal of Science, and the 1989 Nobel Prize in chemistry.

Sean Eddy
Dr. Eddy develops computer algorithms to decipher genomic DNA sequences. He is interested in finding genes that produce catalytic RNAs instead of proteins. Dr. Eddy is an HHMI Assistant Investigator and also Alvin Goldfarb Distinguished Professor of Computational Biology in the Department of Genetics at Washington University School of Medicine, St. Louis. He received his B.S. degree in biology from the California Institute of Technology and his Ph.D. degree in molecular biology from the University of Colorado at Boulder, under the direction of Larry Gold. He worked briefly in industry at NeXagen Pharmaceuticals, followed by postdoctoral work with Richard Durbin and John Sulston at the

MRC Laboratory of Molecular Biology in Cambridge, England, where he began to develop algorithms for computational analysis of genome sequences. He is a coauthor of the book Biological Sequence Analysis: Probabilistic Models of Proteins and Nucleic Acids.

David Eisenberg

David Eisenberg is an Investigator of the Howard Hughes Medical Institute and also serves as Director of the UCLA-DOE Laboratory of Structural Biology and Molecular Medicine at the University of California, Los Angeles. His work involves the determination and analysis of protein structures, with special interest in protein interactions. By combining information from structures, genome sequences, DNA microarrays, and from the scientific literature, Eisenberg is studying the networks of interacting proteins that control the lives of cells. He was elected to membership in the National Academy of Sciences in 1989. He received his Ph.D. from Oxford University, England.

Karen Hersey

Karen Hersey is the former Senior Counsel for Intellectual Property at the Massachusetts Institute of Technology in Cambridge, Massachusetts. Ms. Hersey joined MIT as a technology licensing attorney in 1980, just as technology transfer was becoming an important activity for U.S. research universities. In addition to licensing MIT's patented technology she had primary responsibility for MIT's early efforts to commercially license its computer-related technologies under the new U.S. Copyright law. In 1987, Ms. Hersey left MIT to take up the directorship of technology licensing at North Carolina State University in Raleigh. In 1990, Ms. Hersey returned to MIT as principal legal advisor on intellectual property matters and policy for MIT and with responsibility for advising on computer law and copyright-related issues. This responsibility led to an early interest in the problems universities confront in coping with use of copyrighted materials and the licensing of information products for educational and research use. Ms. Hersey now serves as an Advisor on Intellectual Property to MIT. She holds a bachelor's degree from

Goucher College in Baltimore, Maryland and a law degree from Boston University and is a member of the Massachusetts and North Carolina Bars. An active member of the university technology-transfer community, she has chaired and participated in numerous workshops and seminars on technology transfer practice and intellectual property law, and is a past president of the Association of University Technology Managers. For the past five years, Ms. Hersey has served as a consultant to the Association of Research Libraries (ARL) on intellectual property and other legal issues confronting university libraries. She worked with the ARL to develop both a basic and advanced course for librarians covering copyright and the licensing of electronic resources. She participates as an instructor for both courses several times throughout the year.

Steven H. Holtzman
Steven H. Holtzman is the President and Chief Executive Officer of Infinity Pharmaceuticals, Inc. Previously, he was the Chief Business Officer of Millennium Pharmaceuticals, Inc. in which role his responsibilities included licensing, intellectual property and corporate law, government relations, and public policy. Prior to joining Millennium, from 1986 to 1994, Mr. Holtzman was a founder and the first employee of DNX Corporation, the first commercial enterprise devoted to the development of biomedical and pharmaceutical applications of transgenic animal technology. In the early 1980s, Mr. Holtzman conceived of and was the founding Executive Director of the Ohio Edison Program, the nation's first state government program directed to achieving economic development through funding young technology-based ventures and university/industry collaborative research and development efforts. In the late 1970s, Mr. Holtzman was an instructor and tutor in moral philosophy and the philosophy of language at Corpus Christi College, Oxford University, U.K. In 1995, Mr. Holtzman was a founding Co-Chair and is a current member of the Biotechnology Industry Organization (BIO) Bioethics Committee. In 1998, he served as a member of the Working Group of the Advisory Committee to the Director of the NIH on Access to Research Tools. In 1996, he was appointed by President Clinton as

the sole individual from the pharmaceutical or biotechnology industry to serve on the National Bioethics Advisory Commission, the principal advisory body to the President and Congress on ethical issues in the biomedical and life sciences. In late 1999, he was asked to serve a second term on the Commission. Since 1999, he has served as a Trustee of The Hastings Center for Bioethics. Mr. Holtzman received his B.A. in Philosophy from Michigan State University and his B.Phil. graduate degree in Philosophy from Oxford University, which he attended as a Rhodes Scholar. He is a frequent invited speaker at biotechnology and pharmaceutical industry conferences, NIH and NAS symposia, and business schools on the subjects of structuring alliances between biotechnology and pharmaceutical companies, biotechnology entrepreneurship, bioethics, and patents and intellectual property protection in the life sciences.

George Poste
George Poste is Chief Executive of Health Technology Networks, a consulting group specializing in the application of genomics technologies and computing in health care. From 1992 to 1999 he was Chief Science and Technology Officer and President, Research and Development of SmithKline Beecham (SB). During his tenure at SB he was associated with the successful registration of 29 drug, vaccine, and diagnostic products. Dr. Poste is Chairman of diaDexus and Structural GenomiX in California and serves on the Board of Directors of Maxygen, Illumina and Orchid Biosciences. He is also advisor to several venture capital funds. He is a fellow of Pembroke College Cambridge and Distinguished Fellow at the Hoover Institution and Stanford University. He is a member of the Defense Science Board of the U.S. Department of Defense and in this capacity he chairs the Task Force on Bioterrorism. He is also a member of the National Academy of Sciences Working Group on Defense Against Bioweapons.

Dr. Poste is a Board Certified Pathologist, a Fellow of the Royal Society, the National Academy of Great Britain and a Fellow of the

Academy of Medical Sciences. He has published over 350 scientific papers, coedited 15 books on cancer, biotechnology, and infectious diseases and serves on the editorial board of multiple technical journals. He is invited routinely to be the keynote speaker at a wide variety of academic, corporate, investment, and government meetings to discuss the impact of biotechnology and genetics on health care and the challenges posed by bioterrorism.

Natasha Raikhel

Natasha Raikhel is the Director of the newly organized Center for Plant Cell Biology (CEPCEB) within the Genomics Institute at the University of California at Riverside (UCR). She holds the Ernst and Helen Leibacher Endowed Chair in Plant Molecular, Cell Biology & Genetics, and is also a Distinguished Professor of Plant Cell Biology. Natasha Raikhel received her M.S. in Biology and her Ph.D. from the Institute of Cytology in Leningrad, USSR. She has served on numerous government and industry advisory boards and several editorial boards and was appointed Editor-in-Chief of *Plant Physiology* in May 2000. She was awarded the 2002 Senior Career Recognition Award by the Women in Cell Biology Committee of the American Society for Cell Biology (ASCB) and is recognized as one of the most highly cited researchers in the field of plant science. Prior to working at UCR, Natasha served as Professor in the DOE-Plant Research Laboratory at Michigan State University where she developed a research program to study the plant genes involved in nuclear and vacuolar protein sorting in *Arabidopsis thaliana*. Research in her laboratory is presently focused on understanding the fundamental principles of vacuolar biogenesis and protein trafficking through the secretory system and on elucidation of the components that mediate cell wall biosynthesis in plants. Her multidisciplinary approach utilizes a combination of cellular, molecular, genetic, proteomic, and genomic technologies. She has been working in this field for three decades and has guided many graduate and post-doctoral associates in their research.

Richard Scheller

Richard H. Scheller, Ph.D., joined Genentech in 2001 as senior vice president, Research. Prior to joining Genentech, Scheller served as professor of Molecular and Cellular Physiology and of Biological Sciences at Stanford University Medical Center and as an investigator at the Howard Hughes Medical Institute. Scheller received his first academic appointment to Stanford University in 1982. Scheller's work in cell and molecular biology has earned him numerous awards including the 1997 National Academy of Sciences Award in Molecular Biology. He is a member of the Academy of Arts and Sciences and the National Academy of Sciences and is on the editorial board of several journals including *Neuron, Molecular Biology of the Cell,* and the *Journal of Cell Biology*. Scheller has served on numerous advisory boards including the National Advisory Mental Health Council of the National Institutes of Health. Scheller holds a doctorate in chemistry from the California Institute of Technology where he was also a postdoctoral fellow, Division of Biology. He was also a postdoctoral fellow at Columbia University, College of Physicians & Surgeons.

David Singer

David B. Singer is the Chairman and Chief Executive of GeneSoft, a company focused on developing medical therapies to treat gene-mediated disease. Prior to joining GeneSoft, Mr. Singer was Senior Vice President and Chief Financial Officer of Heartport, Inc. He served as the founding President and Chief Executive Officer of Affymetrix, Inc. Mr. Singer also held business development and finance positions at Affymax NV and Baxter Healthcare Corporation. He is a director of Affymetrix and Corcept, Inc. Mr. Singer received his B.A. in History from Yale University and his M.B.A. from the Graduate School of Business at Stanford University. He is a Henry Crown Fellow of the Aspen Institute.

Mary Waltham

Mary Waltham, most recently President and Publisher for *Nature* and the *Nature* family of journals in the United States, and formerly Managing Director and Publisher of *The Lancet* in the UK, founded her own consulting company two years ago. Its purpose is to help international scientific, technical and medical publishers to confront the rapid change that the networked economy poses to their traditional business models, and to develop new opportunities to build publications that deliver outstanding scientific and economic value. She has worked at a senior level in science and medical publishing companies across a range of media, which includes textbooks, magazines, newsletters, journals, and open learning materials.

APPENDIX B

Community Standards for Sharing Publication-Related Data and Materials

The Lecture Room
National Academy of Sciences
Washington, D.C. 20418

February 25, 2002

AGENDA

8:30 am   Welcome, Orientation for the day.
          Dr. Thomas Cech, Study Chairman

8:45   Keynote Presentation: Dr. Eric Lander,
          The Whitehead Institute for Biomedical Research

9:15   Questions and Answers from audience and webcast listeners.

9:30   Introduction of Situation 1: The Rare Resource.   Dr. Cech

9:35   Panelists Response:

          Maria Freire, CEO, Global Alliance for TB Drug Development (former director of NIH Office of Technology Transfer)

          Elizabeth Neufeld, Professor and Chair, Department of Biological Chemistry, UCLA School of Medicine

Ira Mellman, Editor, *Journal of Cell Biology*, and Professor, Department of Cell Biology, Yale University

Michael Hayden, Professor of Medical Genetics, University of British Columbia, Director, UBC Center for Molecular Medicine and Therapeutics; and Chief Scientific Officer, Xenon Genetics, Inc.

10:30   Plenary Discussion of Situation 1
Do the positions take into account all the realities of the situation and the pressures on an investigator?
Can a consensus be reached?
How might it be implemented?
How to deal with noncompliance?

11:15   Working Groups adjourn to discuss Situations 2a 2b 2c

1:10 pm   Adjourn working groups, Reconvene in Plenary. Webcast continues.

1:15   Reporting by working group rapporteurs.

2a (10 min) Publishing Primary Brain Imaging Data
Rapporteur: Fintan Steele, Editor, *Molecular Therapy*, and Executive Editor, *Genomics*

2b (10 min) Genome Sequences of Quantitative Trait Loci
Rapporteur: Robert Hazelkorn, Fanny L. Pritzker Distinguished Service Professor, Department of Molecular Genetics and Cell Biology, University of Chicago

2c (10 min) Obligations in Exchange for Materials Received
Rapporteur: Corey S. Goodman, Evan Rauch Professor of

Neuroscience, University of California, Berkeley; and President and CEO, Renovis, Inc.

30 min plenary discussion

2:15   Introduction of Situation 3: The Virtual Heart. Dr. Cech

2:20   Panelists Response:

Ari Patrinos, Associate Director, Office of Biological and Environmental Research Office of Science, Department of Energy

James Wells, President and Chief Scientific Officer, Sunesis Pharmaceuticals, Inc.

Bob Waterston, James S. McDonnell Professor and Head, Department of Genetics, Washington University School of Medicine

Barbara Cohen, Editor, *The Journal of Clinical Investigation*

3:00   Questions and Comments from audience and webcast listeners.

4:15   Consensus building: Where is there agreement in principle, and how can the community deal with noncompliance? Dr. Cech

5:00   Adjourn

Workshop Situations

Situation 1. The Rare Resource

Issues raised:

1. **What are an author's obligations to share a reagent on which he or she has published, especially given a rare resource?**

2. **Who is responsible for enforcing the requirements for investigators to share published materials?**

A well-known senior investigator publishes a paper with others in his lab, including two graduate students, in which knockout mice are generated and characterized, in part, with a polyclonal antibody.

The mice reproduce poorly and the investigator plans on using the antibody for many further experiments and only has a limited amount of the reagent. After getting requests for the materials associated with the paper, he decides to write back and suggest he is willing to distribute the reagents as part of a collaboration that would include coauthorship with some of the requesters.

A young investigator also requested the materials but hasn't gotten an answer, even though she has attempted to make contact several times in the last four months. She believes that there is no way the journal would ever sanction this famous investigator and wonders if she should complain to the agency that funded his work. The young investigator is concerned that her career may be damaged by these actions.

Questions for consideration by panelists:

1. What are the pressures faced by the young scientist?
2. What is the best way for the young scientist to get satisfaction?
3. What is a reasonable time to allow someone to turn around a request for a reagent?
4. Under what circumstances is it fair for the senior investigator to request collaboration in exchange for a published material?
5. How would the situation be different if a company had published the paper?

Situation 2a. Primary Brain Imaging Data

Issues raised:

1. What should journal editors do when authors refuse or are hesitant to supply all their primary data to reviewers?

2. To what extent is "data mining" a privilege of the team who originally collects the primary data?

3. To what extent should primary data involving expensive facilities and human or higher animal subjects be made available for evaluation and alternative interpretation? Within what time limits, and what constraints?

4. Is the cost and time it takes to collect data and materials a factor? If so, in which direction?

5. Is the source of funding (private or public) a factor?

6. What principles should guide investigators in this area?

A team funded by NIH submits a short article for publication to a leading journal *Neural Hieroglyphica* which comprises an analysis of functional and physical changes in the brains of schizophrenics, based on the results of several detailed fMRI-imaging procedures. Note: fMRI studies are always based on taking differences between images; only the 'difference image' [the subtraction of an image taken under one circumstance from an image taken in another circumstance] is normally published, and its appearance depends notoriously on how thresholds are taken.

Point 1: One of the reviewers approached by the editor asks the journal staff to obtain the complete set of fMRI data from the authors, as she wants to check through some of the findings reported in the paper. The authors refuse. Should the journal press the authors (and state that if peer review cannot be completed on this paper as and by reviewers selected by the journal then there is no interest in publishing it)? OR find another reviewer?

After publication of the short article, a colleague asks to see the primary data which the authors had used to construct the images (only one 'typical image' was published) on which measurements were taken. The published data consisted of a summary of the quantitative measurements.

The colleague is in a department of psychology at an institution without fMRI facilities and without a medical school, and is in no position to acquire similar primary data. He finds that the conclusions to the short article do not fit with his anatomical studies of the brains of deceased schizophrenics, and wonders if there is an alternate interpretation of the primary data.

The team claims that the data is being used in a follow-on paper they are writing and declines to share the data at this moment. They also note that it took them 2 years to identify the subjects and collect the data, and think that they should be able to "mine" it before others do. Several postdocs are said to be pinning their futures on getting papers out of this data. (However, the paper was published in *Neural Hieroglyphica*, and on average, only half of the short papers published in this journal have timely longer articles written to back them up.)

Point 2: Is the primary investigator acting within his/her rights?

Situation 2b. Gene Sequences for Quantitative Trait Loci

Issues raised:

1. Does withholding data compromise the quality of the results the company wants to publish?

2. Is the confidential information essential to support and understand the data?

3. Can the results without the confidential information benefit and accelerate public research?

AgriGenome, a plant genomics company, is studying the quantitative trait loci (QTL) governing drought resistance in wheat. They have identified a QTL down to 100 kb. They have done sample sequencing and have fragments of genes that appear promising with respect to function, but they haven't pinned down a single gene responsible for drought resistance yet. Importantly, they have data showing that this QTL alone is able to increase drought resistance.

They submit a manuscript in which they show that this single QTL is able to increase drought resistance. They report detailed biological studies, including field tests under conditions of abundant water, limiting water, and extreme drought that show a strong correlation of drought resistance with the presence of markers for the QTL in different germplasm. Because they haven't pinned down a gene yet, they do not want to publish any of the sample sequencing data. They have just filed for a patent on the markers, but because those could easily be used to find linked markers that would not be covered under the patent, they wish not to publish the sequence of the markers.

One reviewer wants to require that both the marker sequences and the DNA sequence of at least one gene be described in the paper and deposited in the database upon publication, while the second reviewer requires only the sequences of the markers. The journal would like to publish the paper, because they would be the first journal with a paper showing that this important trait is at a single QTL, and the journal is trying to establish itself in plant genomics. As written, it would be difficult, but not impossible, for others to reproduce the work; they would have to obtain the original lines (which would be in the public domain) and do several cycles of crosses to recreate germplasm from which, following the procedures in the paper, the markers could be isolated. It might take 3 to 4 years, versus one field season if they had the marker sequences. What should the editor decide?

Situation 2c. Obligations in Exchange for Materials Received

Issues raised:

1. Is it appropriate for materials providers to require a right to review publications? If so, at what point should the review take place? Prior to submission for publication, prior to actual publication? Should the materials provider have the right to require changes to the publication?

2. Should the right of review be enforceable? What recourse is appropriate if publication review obligations are ignored by the investigator?

3. What are the pros and cons a company should consider before releasing materials for use in research that will be published? Does the investigator have a responsibility to consider the provider's interests in its materials?

The Gendefex company is a small biotech company that wants to allow its investigators to publish, and is willing to give academic investigators access to materials referred to in its investigators' publications with a material transfer agreement (MTA). It recently received requests for two different reagents that were described in recent articles published by Gendefex biologists.

In the case of the first reagent, the proposed MTA includes a provision that would allow Gendefex the right to use improvements to the reagent made by the requesting investigator, and give the company the "freedom to operate" should the requesting investigator find new, potentially patentable, uses for the material.

In the second case, the MTA has no special provisions about the rights of Gendefex related to the investigators' use of the reagent.

However, in both cases, the MTA stipulates that Gendefex should receive copies of manuscripts that include information related to the use of its materials 60 days prior to submission for publication. The university officials of one investigator who wants the materials are unwilling to sign the MTA with this requirement because it delays the publication process.

The president of the company, although a former academic, himself, is frustrated that some investigators who use the materials do not honor the request for the manuscript even though they sign the agreement and is disinclined to relent on the requirement. It makes him angry that he feels he has no recourse against noncompliant investigators and feels his company is being used.

Lawyers in the company are against permitting company investigators to publish if it means handing over materials for academic research. They point out that publications they have reviewed prior to publication describe a plethora of discoveries surrounding the company's materials and reveal information the company does not want its competitors to have. In addition, some discoveries are in areas the company was working on and intended to patent itself. Because of materials made available to the university investigators, the company will have to bargain for licenses to intellectual property it needs but does not hold.

Questions for consideration by panelists:

1. What is the motivation for Gendefex to publish?
2. In each case, is the request to review copies of manuscripts prior to submission for publication reasonable or necessary? Is the request enforceable?
3. What obligations do the academic community have to the for-profit sector if they acquire commercial materials for research? Are there reasonable mechanisms that can be utilized to protect the company's interests?

4. How does or how should a company evaluate the pluses and minuses of publishing in academic journals?
5. What is the reality of the "timing" of publications in the academic and for-profit sectors?

Situation 3. The Virtual Heart

Issues raised:

1. Review: How should increasingly complex software and databases be evaluated for publication? What depth of materials must be provided to reviewers?

2. Repeatability: Science relies on open disclosure of results in publications, and the possibility of replication (or refutation) of the results by other scientists. What information must be disclosed to enable repeatability? Does repeatability of science that involves complicated software or databases require complete access to computer source code and raw data, or only the possibility of running the code and interrogating the database?

3. Responsibility of the authors to enable future work: The end purpose of a scientific publication is to describe a result that others in the scientific community can build on. A paper describing a new fact or observation is generally sufficient to suggest new ideas and experiments to other investigators. However, an increasing number of papers describe large resources that are difficult to reproduce, such as a database or a software package. Should such resources be made freely available to the community as a condition of publication? If not, what is the societal purpose of scientific papers describing resources that are proprietary products? Certain types of archival data, such as genome sequences or protein structure coordinates, may be generated in the course of a "standard" result-oriented paper, but these data, if available, may be mined in ways entirely unexpected by the original investigators; when should such data be deposited in public repositories, and when, if ever, should deposition be mandatory?

4. **Enforcement:** Who enforces openness? Is this the job of funding agencies; if so, should private companies be allowed to play by different rules? Is it the job of journals; if so, what mechanism prevents backsliding in the case of a "hot" paper that a journal has a strong desire to publish for its own reasons of profitability and publicity?

In 2015, the Cardiomics company announces that its proprietary model of the human heart (Virtual Heart) is complete, including heart models reflecting various stages of cardiovascular disease, genetic disorders, and the consequences of infarction. Virtual Heart incorporates an extensive database of experimental data collected by Cardiomics. Cardiomics is basing its impending IPO on Virtual Heart and is trying to gain publicity about it in the media. As part of this effort Cardiomics submits two papers on Virtual Heart to high-profile, for-profit journals.

The proprietary Cardiomics database is 500 terabytes of Oracle relational database tables. Among its many components are data on genetic polymorphism profiles in families with heart disease, MRI images, and electrocardiograms. In 2015, widely recognized public data repositories exist for human genetic profiles (PolymorphismDB) and MRI images (MRI-DB) but not for ECGs. The Virtual Heart program itself is 500,000 lines of code.

Paper A gives an overview of the entire Virtual Heart system, including the software and the database. The central point of this paper is that Virtual Heart is a useful system for virtual cardiac experimentation and diagnosis. Neither the database nor the software are available from Virtual Heart; they are closely held proprietary assets.

Paper B describes a specific result in which the Virtual Heart system is used to predict that thrombospondin variants are probably associated with early heart attacks. This computational prediction is validated by experimental results that are fully described in the paper; one SNP in a thrombospondin gene is shown to be associated with heart disease in family pedigrees. The central point of the paper is that this thrombospondin variant will be useful for diagnostics and screening.

Questions for consideration by Panelists:

1. What materials should the reviewers receive in order to evaluate these manuscripts? Does it make a difference whether the point of the paper is Virtual Heart itself, or an independently verified result obtained from the system?

2. What data should be made available to the scientific community upon publication? Does this differ for the two papers? Does it matter whether public data repositories exist or not? What restrictions should Cardiomics be allowed to put on any data release to protect its proprietary interests? Is it acceptable if the data are made available, but only under expensive proprietary licensing terms?

3. What are the obligations of the journals to the scientific community? Would the obligations differ for journals that are society-based as opposed to for-profit? If a journal's editorial decisions are inconsistent with scientific community standards, how should the scientific community respond?

4. What restrictions should Cardiomics be allowed to put on any data release to protect its proprietary interests? Would this be any different for a large pharmaceutical company with less at risk than a small bioinformation company like Cardiomics—or for an academic lab trying to protect itself from academic competition?